# Nuclear Terrorism and Global Security

This book examines the prospects and challenges of a global phase-out of highly enriched uranium – and the risks of this material otherwise being used by terrorists to make atom bombs.

Terrorist groups, such as Al Qaeda, have demonstrated repeatedly that they seek to acquire nuclear weapons. Unbeknownst even to many security specialists, tons of bomb-grade uranium are trafficked legally each year for ostensibly peaceful purposes. If terrorists obtained even a tiny fraction of this bomb-grade uranium they could potentially construct a nuclear weapon like the one dropped on Hiroshima that killed tens of thousands.

Nuclear experts and policymakers have long known of this danger but – so far – have taken only marginal steps to address it. This volume begins by highlighting the lessons of past successes where bomb-grade uranium commerce has been eliminated, such as from Argentina's manufacture of medical isotopes. It then explores the major challenges that still lie ahead: for example, Russia's continued use of highly enriched uranium (HEU) in dozens of nuclear facilities. Each of the book's thirteen case studies offers advice for reducing HEU in a specific sector. These insights are then amalgamated into nine concrete policy recommendations for U.S. and world leaders to promote a global phase-out of bomb-grade uranium.

This book will be of much interest to students of nuclear proliferation, global governance, international relations, and security studies.

**Alan J. Kuperman** is Associate Professor at the LBJ School of Public Affairs, University of Texas at Austin, where he is also Coordinator of the Nuclear Proliferation Prevention Project (www.nppp.org).

## Routledge Global Security Studies

Series Editors: Aaron Karp and Regina Karp

**Nuclear Proliferation and International Security**
*Edited by Morten Bremer Maerli and Sverre Lodgaard*

**Global Insurgency and the Future of Armed Conflict**
Debating fourth-generation warfare
*Terry Terriff, Aaron Karp and Regina Karp*

**Terrorism and Weapons of Mass Destruction**
Responding to the challenge
*Edited by Ian Bellany*

**Globalization and WMD Proliferation**
Terrorism, transnational networks, and international security
*Edited by James A. Russell and Jim J. Wirtz*

**Power Shifts, Strategy, and War**
Declining states and international conflict
*Dong Sun Lee*

**Energy Security and Global Politics**
The militarization of resource management
*Edited by Daniel Moran and James A. Russell*

**US Nuclear Weapons Policy after the Cold War**
Russians, 'rogues' and domestic division
*Nick Ritchie*

**Security and Post-Conflict Reconstruction**
Dealing with fighters in the aftermath of war
*Edited Robert Muggah*

**Network Centric Warfare and Coalition Operations**
The new military operating system
*Paul T. Mitchell*

**American Foreign Policy and the Politics of Fear**
Threat inflation since 9/11
*Edited by A. Trevor Thrall and Jane K. Cramer*

**Risk, Global Governance and Security**
The other war on terror
*Yee-Kuang Heng and Kenneth McDonagh*

**Nuclear Weapons and Cooperative Security in the 21st Century**
The new disorder
*Stephen J. Cimbala*

**Political Economy and Grand Strategy**
A neoclassical realist view
*Mark R. Brawley*

**Iran and Nuclear Weapons**
Protracted conflict and proliferation
*Saira Khan*

**US Strategy in Africa**
AFRICOM, terrorism and security challenges
*Edited by David J. Francis*

**Great Powers and Strategic Stability in the 21st Century**
Competing visions of world order
*Edited by Graeme P. Herd*

**The Globalisation of NATO**
Intervention, security and identity
*Veronica M. Kitchen*

**International Conflict in the Asia-Pacific**
Patterns, consequences and management
*Jacob Bercovitch and Mikio Oishi*

**Nuclear Proliferation and International Order**
Challenges to the Non-Proliferation Treaty
*Edited by Olav Njølstad*

**Nuclear Disarmament and Non-Proliferation**
Towards a nuclear-weapon-free world?
*Sverre Lodgaard*

**Nuclear Energy and Global Governance**
Ensuring safety, security and non-proliferation
*Trevor Findlay*

**Unipolarity and World Politics**
A theory and its implications
*Birthe Hansen*

**Disarmament Diplomacy and Human Security**
Regimes, norms and moral progress in international relations
*Denise Garcia*

**Causes and Consequences of Nuclear Proliferation**
*Edited by Robert Rauchhaus, Matthew Kroenig and Erik Gartzke*

**Why Did the United States Invade Iraq?**
*Edited by Jane K. Cramer and A. Trevor Thrall*

**Regional Powers and Security Orders**
A theoretical framework
*Edited by Robert Stewart-Ingersoll and Derrick Frazier*

**A Perpetual Menace**
Nuclear weapons and international order
*William Walker*

**Iran's Nuclear Programme**
Strategic implications
*Joachim Krause*

**Arms Control and Missile Proliferation in the Middle East**
*Edited by Bernd Kubbig*

**The National Politics of Nuclear Power**
Economics, security and governance
*Benjamin Sovacool and Scott Valentine*

**Arms Controls in the 21st Century**
Between coercion and cooperation
*Edited by Oliver Meier and Christopher Daase*

**Reconceptualising Deterrence**
Nudging toward rationality in Middle Eastern rivalries
*Elli Lieberman*

**Psychology, Strategy and Conflict**
Perceptions of insecurity in International Relations
*Edited by James W. Davis*

**Nuclear Terrorism and Global Security**
The challenge of phasing out highly enriched uranium
*Edited by Alan J. Kuperman*

**Ballistic Missile Defence and US National Security Policy**
Normalisation and acceptance after the Cold War
*Andrew Futter*

# Nuclear Terrorism and Global Security

## The challenge of phasing out highly enriched uranium

**Edited by**
**Alan J. Kuperman**

Routledge
Taylor & Francis Group
LONDON AND NEW YORK

First published 2013
by Routledge
2 Park Square, Milton Park, Abingdon, Oxon, OX14 4RN

Simultaneously published in the USA and Canada
by Routledge
711 Third Avenue, New York, NY 10017

*Routledge is an imprint of the Taylor & Francis Group, an informa business*



*British Library Cataloguing in Publication Data*
A catalogue record for this book is available from the British Library

*Library of Congress Cataloging in Publication Data*
Nuclear terrorism and global security : the challenge of phasing out highly enriched uranium / edited by Alan J. Kuperman.
pages cm. -- (Routledge global security studies) Includes bibliographical references and index.
1. Uranium industry--Security measures. 2. Highly enriched uranium--Government policy. 3. Uranium enrichment--Government policy. 4. Terrorism--Prevention. 5. Security, International.
I. Kuperman, Alan J., editor of compilation.
HD9539.U7N83 2013
363.325'5--dc23
2012039641

ISBN: 978-0-415-66068-6 (hbk)
ISBN: 978-0-203-07390-2 (ebk)

Typeset in Times New Roman
by Saxon Graphics Ltd., Derby

MIX
Paper from responsible sources
FSC® C013604

**This book is dedicated to the memory of**
**Paul L. Leventhal**

**“I don’t want to unduly alarm. But to duly alarm.”**

# Contents

# List of Figures and Tables

## Figures

## Tables

# Notes on Contributors

**Jared Berenter** is completing a dual Master's Degree in Global Policy Studies and Latin American Studies at the University of Texas at Austin. He specializes in environmental policy and international security, with a focus on food security, natural disaster risk reduction, and crisis management in developing countries.

**Braden Civins** has a J.D. from The University of Texas School of Law and a Master of Global Policy Studies from the LBJ School of Public Affairs.

**Chloe Colby** is a writer and editor at the private intelligence firm Stratfor. She holds a Master of Global Policy Studies degree from the University of Texas at Austin's LBJ School of Public Affairs. She was awarded a Foreign Languages and Area Studies Fellowship in 2010 to study the German language and European security issues.

**Kenneth Dayman** is a Ph.D. student in the Nuclear and Radiation Engineering program at the University of Texas at Austin. His research interests include novel radiation detectors, systems, and data analysis, especially as applied to nuclear safeguards and nuclear forensics technologies.

**Christine Egnatuk** has a Ph.D. in Nuclear and Radiation Engineering from the University of Texas at Austin. She has been a DoE/NNSA Nuclear Nonproliferation and International Safeguards Fellow. Her research interests include noble gas detection and PGAA systems.

**Alexander Fay** is a DOE/NNSA Nuclear Nonproliferation International Safeguards Fellow, and a Ph.D. student in the Nuclear and Radiation Engineering program at the University of Texas at Austin, where he also earned his M.S. His current research focuses on the quantification of radio-argon background source terms for verification of the Comprehensive Nuclear-Test-Ban Treaty.

**Shing-yao (Sandra) Feng** holds a Master of Global Policy Studies degree from the LBJ School of Public Affairs at The University of Texas at Austin. She has received a Fulbright Fellowship.

**Margarita Jimenez** holds a Master of Global Policy Studies degree from the LBJ School of Public Affairs at the University of Texas at Austin, where she was Treasurer for the Public Affairs Alliance for Communities of Color and co-founder of the Student Veterans Association. She also received the McNair Fellowship.

**Alan J. Kuperman** is Associate Professor at the LBJ School of Public Affairs, University of Texas at Austin, where he is also Coordinator of the Nuclear Proliferation Prevention Project (www.nppp.org). He holds a Ph.D. in Political Science from the Massachusetts Institute of Technology, an M.A. in International Relations from Johns Hopkins University, and an A.B. in Physical Sciences from Harvard University.

**R. Blake Messer** recently earned a Master of Global Policy Studies degree from the LBJ School of Public Affairs. He specializes in Global Technology Policy, focusing on Internet policy, intellectual property, energy, security, and the digital divide. He has been a William H. Crook Fellow of International Affairs, working in Peru.

**Paul Osborne** recently graduated with a dual Master's degree from The University of Texas at Austin in Global Policy Studies and Russian, East European, Eurasian Studies. He specializes in security, law, and diplomacy issues.

**Yaroslav Primachenko** recently graduated from The University of Texas at Austin with a Master of Global Policy Studies degree from the LBJ School of Public Affairs. His policy interests include national security, global stability, and international trade and commerce. He was born in Kiev, Ukraine, and moved to the United States in 1995.

**Kendra Vessels** has a dual Master's degree from The University of Texas at Austin in Global Policy Studies and Russian, East European, Eurasian Studies. She previously served in the US Air Force.

**Rebecca Ward** is a Stanton Nuclear Security Fellow at Harvard University, during 2012–2013. She is a Ph.D. candidate in the Nuclear and Radiation Engineering Program at the University of Texas at Austin, supported by the DHS/DoD Nuclear Forensics Graduate Fellowship. Her research focuses on optimization of the safeguards regime. She has previously worked at three national laboratories: Pacific Northwest, Brookhaven, and Argonne.

# Acknowledgements

After working to reduce the dangers of highly enriched uranium (HEU) for more than a quarter-century – in Congress, academia, and a non-governmental organization – I have too many people to thank, and so will single out only a few, apologizing sincerely to the rest.

Paul L. Leventhal handed me a thick stack of papers in early 1987, at the non-governmental Nuclear Control Institute (NCI), where I had just started my first job after college. He told me to read them so that I could begin advocating for federal funding to convert U.S. university nuclear reactors from bomb-grade HEU fuel to a low-enriched alternative that would reduce risks of nuclear terrorism. Thus began a two-decade mentorship in nuclear affairs, investigative research, press relations, writing, lobbying, and professional conduct – for which I forever will be indebted. When Paul fell terminally ill, he arranged for NCI's archives and endowment to be transferred to the University of Texas at Austin, so that I could establish the Nuclear Proliferation Prevention Project (NPPP). This book represents the NPPP's first major publication.

Soon after Paul's passing, I invited Frank von Hippel to Austin, TX. We too had encountered each other originally in 1987, at my first meeting of a coalition dedicated to reducing weapons-usable nuclear materials. Twenty years later, as we walked along Town Lake, I asked if he could suggest a first research project for the NPPP. He recommended that I expand my past work on phasing out HEU in research reactors to encompass all worldwide non-weapons uses of HEU. And so I did. He is thus the intellectual father of this volume.

To address the broader agenda, I needed to deploy 13 graduate students to five continents, and that required money. But all of the potential funders that I asked rejected me, except one, the non-governmental Nuclear Threat Initiative. I thank NTI's vice-president Page Stoutland, and former vice-president Laura Holgate, for their financial support of – and faith in – this NPPP initiative.

Transforming research papers into a book manuscript also required funding, including for travel to vet the preliminary findings with experts worldwide. For the grant to enable that, I thank the Policy Research Institute of the LBJ School of Public Affairs, University of Texas at Austin, and Associate Dean Robert Wilson.

The major reason that we can today discuss a prospective global HEU phase-out is because of the groundwork laid since 1978 by two U.S.-led, international initiatives to eliminate HEU from research reactors and medical-isotope production. The leaders of both efforts graciously visited Austin to offer our researchers vital background information and then – along with other experts – provided substantial help during their investigations. My special thanks to Jim Matos, Parrish Staples, and Armando Travelli.

My colleague at UT-Austin, Prof. Steven Biegalski, Director of the university's Nuclear Engineering Teaching Laboratory, was indispensable. He encouraged his nuclear engineering Ph.D. students to participate in this year-long policy project, delivered a guest lecture, and hosted us on a guided tour of his facility's nuclear reactor.

The relatively young authors who researched and wrote this book exceeded every expectation. Combining nuclear engineers and policy wonks turned out to produce great synergies.

The editors at Routledge took a chance on a topic that no previous book had addressed, because they thought it was important and well done. Thank you to Andrew Humphrys, Regina Karp, and Andrew Karp.

Last but not least, I must thank the other NCI boys – Tom Clements, Steven Dolley, Daniel Horner, and Edwin Lyman – who all still work on nuclear affairs and have been a great source of help and inspiration over the years. Paul would be proud.

# Part I

# Introduction

# 1 Global HEU Phase-Out

## Prospects and Challenges

*Alan J. Kuperman*

Global commerce in nuclear weapons-usable, highly enriched uranium (HEU) – ostensibly for non-weapons purposes – poses significant risks of nuclear terrorism and nuclear proliferation.[1] The international community first recognized these dangers in the 1970s and, ever since, has taken progressive action to minimize and secure such trade. Much has been accomplished, but much more could be done, given that the vast majority of non-weapons HEU commerce persists. This ongoing usage also hinders adoption and implementation of a major international arms-control agreement – the Fissile Material Cutoff Treaty (FMCT) – intended to cap nuclear-weapons arsenals by prohibiting the production of HEU (and plutonium) for weapons. So long as states may claim to require additional HEU for non-weapons purposes, which could be diverted to weapons, the FMCT has a loophole that undermines its intent and thereby its attractiveness to some potential signatories. For all these reasons, this book explores the prospects and challenges of a total, global phase-out of HEU commerce.

This introductory chapter starts by reviewing the dangers of HEU commerce and the international steps already taken towards a phase-out. Next, it provides an overview of remaining global, non-weapons HEU use. Following that, it explains the scope, organization, and methodology of the book's studies, and summarizes the findings. The chapter closes with policy recommendations to promote and accelerate a global HEU phase-out.

## Past Efforts: Commendable But Inadequate

The HEU used in non-weapons applications is often identical to, and in some cases even more optimized for nuclear weapons, than that contained in military arsenals. By definition, "HEU" is enriched to at least 20 percent in the fissile isotope U-235 that sustains a chain reaction. As indicated in Figure 1.1, this distinguishes it from natural uranium, which contains less than one percent U-235, and from the low-enriched uranium (LEU) fuel for nuclear energy plants that typically is enriched to less than 5 percent. (Enrichment is a process that preferentially selects the fissile isotope U-235 from the rest of the uranium, which is composed mainly of U-238 that hinders a chain reaction.) The Hiroshima atom bomb utilized uranium enriched to an average of 80 percent.

The HEU in most of today's nuclear weapons has enrichment of 90 to 93 percent. The HEU for non-weapons applications is typically drawn from identical stocks and so has the same weapons-grade enrichment, whether used as fuel for nuclear research reactors, in targets to produce medical isotopes, in critical assemblies to model nuclear reactors, or for other purposes. In at least one application, the fuel for U.S. (and reportedly also UK) naval propulsion reactors, the non-weapons HEU has been enriched to 97 percent, even higher than typical for nuclear weapons. (As the stockpile of such "super-grade" HEU is exhausted, these naval reactors will be converted to fuel of weapons-grade HEU, which ironically is slightly less enriched.)

Compared to the other main fissile material in military arsenals, plutonium, HEU is much easier for terrorists or states to make into a nuclear weapon. Its low spontaneous neutron-emission rate permits a gun-type design – much simpler than the implosion design required for plutonium – and its low radiation level represents little danger to those handling it. The resulting threat was famously summed up by Manhattan Project physicist Luis Alvarez in his memoirs:

> With modern weapons-grade uranium … terrorists, if they had such material, would have a good chance of setting off a high-yield explosion simply by dropping one half of the material onto the other half … Even a high school student could make a bomb in short order.[2]

Hundreds of nuclear weapons could be made with the HEU used today for non-weapons purposes. The amount required for one bomb depends on several factors including the enrichment level and weapon design. The simplest design, a gun type-weapon, works by slamming two sub-critical masses together to form a super-critical mass. A "reflector" around the outside of the assembly reduces the critical mass by reflecting back in neutrons, which otherwise would escape, to help sustain the chain reaction.

A 1998 study by Los Alamos National Laboratory reports the critical mass of various HEU spheres, as summarized in Figure 1.2. *It indicates that the critical mass is less than 20 kg for 94%-enriched HEU surrounded by a 4-inch reflector of natural uranium.*[3] This critical mass could be reduced by use of a more sophisticated reflector, but up to two critical masses would be required to optimize the yield of a gun-type weapon. Accordingly, 20 kg of HEU may be considered

| *Type of Uranium* | *Typical Enrichment* |
|---|---|
| Natural | 0.7% |
| LEU fuel for nuclear powerplants | < 5% |
| HEU in Nuclear Weapons | > 90% |
| Non-Weapons HEU fuel for reactors | > 90% |
| Non-Weapons HEU targets for medical isotopes | > 90% |

*Figure 1.1* Non-Weapons HEU = Weapons HEU

| *Reflector Thickness (in)* | *Extrapolated Critical Mass (kg U)* |
|---|---|
| 3.925 | 19.74 |
| 1.761 | 26.45 |
| 0.695 | 36.41 |

*Figure 1.2* Critical Mass of 94%-Enriched HEU Sphere with Natural Uranium Reflector

Source: Los Alamos National Laboratory

the approximate minimum required for a non-state actor to manufacture a nuclear weapon of simple design. By contrast, the International Atomic Energy Agency (IAEA) defines a "Significant Quantity" as "the approximate amount of nuclear material for which the possibility of manufacturing a nuclear explosive device cannot be excluded." For HEU, it uses the more conservative estimate of 25 kg of U-235, an arbitrary round number, equivalent to 27 kg of HEU enriched to 94 percent.[4]

If HEU were used in an implosion weapon, which reduces the critical mass by compressing the uranium sphere, the amount required could be considerably less, depending on the extent of compression. But policy analysts often arbitrarily pick the round number of 25 kg of HEU as the minimum for a nuclear weapon. By this standard, a metric ton of HEU is sufficient for 40 nuclear weapons.

Starting in the 1950s, the United States and the Soviet Union exported dozens of nuclear research reactors. The United States originally supplied them with LEU fuel in developing countries, but after some technical difficulties with that fuel, switched to exporting HEU fuel for virtually all the research reactors it supplied. By the mid-1960s, annual HEU exports by the United States alone approached three tons, sufficient for more than 100 bombs (see Figure 1.3).

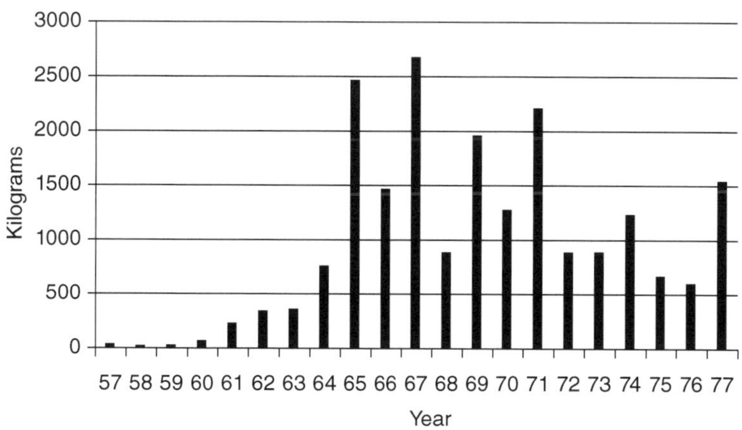


*Figure 1.3* U.S. HEU Exports: 1950s–1970s

In the mid-1970s, following India's misuse of "peaceful" plutonium to manufacture and test a nuclear weapon, the world belatedly awoke to the dangers of commerce in fissile materials. The United States spearheaded an international effort to reduce HEU commerce by converting reactors to fuel of LEU, enriched below 20 percent, which is deemed unsuitable for nuclear weapons because the critical mass is so high. As indicated in Figure 1.4, reducing enrichment from 93 to 19.9 percent can increase the critical mass by nearly 20 times, depending on various assumptions.

The international Reduced Enrichment for Research and Test Reactors (RERTR) program, based at the U.S. Argonne National Laboratory, facilitated this conversion by developing higher-density LEU fuel. As illustrated in Figure 1.5, by increasing the density of the uranium in the fuel, the program was able to sustain the concentration of U-235 necessary for reactor performance, while reducing enrichment from HEU to LEU. Development of increasingly high density fuel over the last three decades has enabled conversion to LEU of progressively higher power research reactors.

RERTR has facilitated a very gradual reduction of global commerce in non-weapons HEU. Established in 1978, the program not only develops higher density LEU fuel but also conducts neutronic and safety calculations and experiments necessary to persuade operators of reactors in the United States and elsewhere that their facilities can operate with LEU fuel and/or produce medical isotopes with LEU targets. In the 1980s, the USSR established an analogous program to convert Soviet-supplied research reactors to lower enriched fuel. In 1986, the U.S. Nuclear

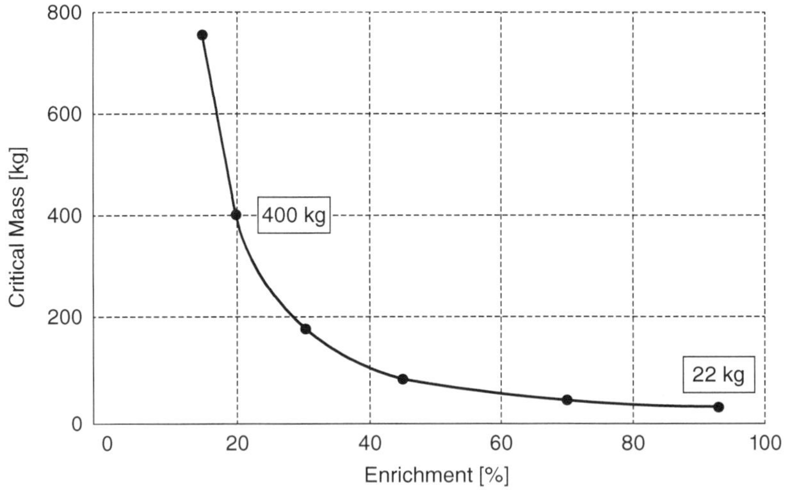


*Figure 1.4 Switch to LEU Fuel Inhibits Misuse for Weapons by Increasing Critical Mass*

Note: Critical mass of a uranium sphere surrounded by a 5-cm beryllium reflector.

Source: Alexander Glaser and Frank N. von Hippel, "Global Cleanout: Reducing the Threat of HEU-Fueled Nuclear Terrorism," *Arms Control Today* (January/February 2006): 18–23.

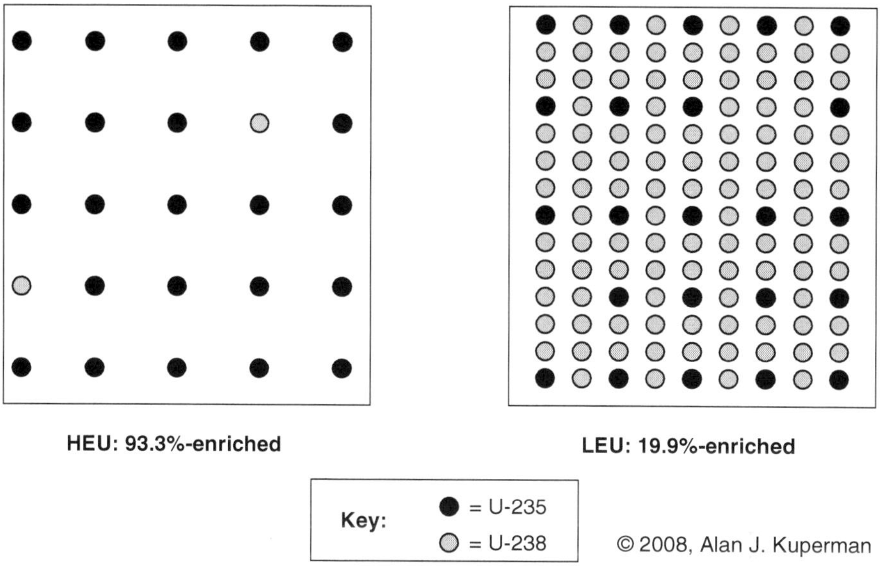


*Figure 1.5* RERTR: Higher Uranium Density Enables Lower Enrichment

Regulatory Commission ordered research reactors under its purview – all domestic ones excluding those of the Department of Energy (DOE) and Pentagon – to convert to LEU fuel as soon as possible. In 2004, the RERTR program was subsumed into the new U.S. Global Threat Reduction Initiative (GTRI), which committed to convert DOE reactors to LEU as well.

As a result of these efforts, nearly all new research reactors built in the world since 1980 use LEU fuel. Moreover, as of October 2012, 83 older research reactors worldwide had been converted to LEU fuel or shut down, of the approximately 200 currently within GTRI's scope. Additionally, the radioisotope that is the basis for most of the world's nuclear medical procedures, Molybdenum-99 (Mo-99), is now produced without HEU in Australia and Argentina – and South Africa, Belgium, and The Netherlands are following suit.

This gradual decline in HEU commerce has helped reduce somewhat the risk that nominally non-weapons HEU will be stolen or diverted for weapons. The overall risk is a function of many variables, including the amount of HEU used, the number of locations at which it is used or stored, the security of those locations, the frequency of transportation, and the security of transportation. RERTR and GTRI have made improvements across all those realms.

## HEU Commerce Persists

To date, however, global efforts to phase out non-weapons HEU commerce have suffered from two major shortcomings. First, they traditionally have focused only on some sectors of HEU commerce, while ignoring others that entail even greater quantities of HEU. Second, even in the sectors of traditional focus, the programs have achieved only partial success.

Past efforts have concentrated on phasing out HEU as fuel for research reactors and targets for medical isotope production, but have faced two obstacles. First, the world's three largest medical isotope producers, responsible for 90 percent of global supply, have resisted conversion to LEU even though it is technically feasible, and instead continue to require approximately a combined 50 kg of HEU annually.[5] Second, the high-performance research reactors that are the most difficult to convert, because they require ultra-high density LEU fuel that has yet to be fully qualified, also require the most fresh HEU fuel. Thus, the RERTR program has converted or shut down mainly the reactors that required the least HEU commerce. Estimates are that the dozens of conversions to date have reduced annual commerce by only 280 kg HEU, while shutdowns have eliminated another 450 kg of yearly use. Meanwhile, unconverted reactors still require an estimated total of 750 kg of fresh HEU annually, mainly at 13 high-performance facilities in the United States, Europe, and the former Soviet Union, as indicated in Figure 1.6.[6]

The second failing of past phase-out efforts has been virtually to ignore some sectors that entail enormous usage of HEU.[7] For example, the RERTR program has very rarely addressed "critical assemblies," which are zero-power facilities used to design reactors by mocking up potential configurations of their fuel.[8] Although they have "lifetime cores," and therefore do not require fresh supplies of HEU, they are tempting targets for wannabe bomb-makers because they typically contain large quantities of HEU in small, lightly irradiated discs, which are relatively easy to handle, transport, and fabricate into weapons. Mainly located in Russia, these facilities together contain an estimated 10 tons of HEU, much more than in all the world's research reactors. (The critical assemblies with the largest HEU inventories are mockups of cores of fast breeder and naval reactors.) Another sector largely ignored by past phase-out efforts has been pulsed reactors, used to simulate the effects of nuclear-weapons detonations, by bringing HEU into a supercritical configuration extremely briefly to produce a burst of neutrons. Such fuel is also attractive for misuse because its large quantity of HEU is only lightly irradiated, leaving it very highly enriched and relatively easy to handle. Pulsed reactors, also located mainly in Russia, together contain more than two tons of HEU.

| | USA | Europe | Russia et al. |
|---|---|---|---|
| | ATR | HFR–ILL | SM–3 |
| | HFIR | FRM–II | MIR.M1 |
| | MURR | BR–2 | WWR–M |
| | NBSR | Orphée | IVV–2M |
| | MIT | | Etc. |
| ***HEU (kgs/yr)*** | ***250*** | ***130*** | ***300+*** |

*Figure 1.6* High-Performance Research Reactors Still Require Large Amounts of HEU Fuel

Naval propulsion reactors in the United States, Russia, the United Kingdom, and India are also beyond the scope of traditional phase-out efforts, but their annual requirement of three tons of HEU fuel is four times that of the world's research reactors. Also excluded from past efforts are the roughly 150 kg of HEU currently used annually in the propulsion reactors of Russian icebreaker ships, which are used to keep open Arctic sea lanes in winter. Finally, phase-out efforts have ignored the HEU used to fuel three-dozen, space-based nuclear reactors launched prior to 1989. No further space reactors have been launched since then, but potential new missions are being contemplated, still based on HEU fuel.

In 2009, DOE announced an expansion of the RERTR scope "from 129 research reactors to 200 research [SIC] reactors."[9] In fact, this broader scope included several other types of facilities discussed above – including icebreaker propulsion reactors, critical assemblies, and pulsed reactors – according to an unpublished fact sheet from the 2009 annual RERTR conference. The additional 71 reactors included 59 in Russia. In 2011, DOE revised this list of newly included facilities to delete 19 that it believed met at least one of three conditions – already shut down, exclusively military, or already using LEU fuel – and added 14 more facilities that it believed did not meet any of those conditions. The net result was to decrease from 71 to 66 the operating facilities that had been added to the RERTR scope, but to increase by five the number of facilities listed as converted or shut down, resulting in no net change to the total number of 200 facilities within the recently expanded RERTR scope.[10] Following the initial announcement in 2009, DOE does not appear ever to have detailed in public the expanded scope of RERTR, at least as of early 2012, which suggests that it is a politically sensitive topic.

Overall, continuing worldwide HEU commerce entails approximately four tons of fresh fuel annually and 12 tons in lifetime cores. As illustrated in Figure 1.7, this considerable, ongoing use of HEU dwarfs the relatively small amounts

| *Nuclear Activity* | *Kgs Used per Year* | *Kgs in Lifetime Cores* |
|---|---|---|
| Research Reactors | 750 | 100s |
| Medical Isotope Targets | 50 | |
| Naval Propulsion | 3,000 | |
| Icebreaker Ship Propulsion | ~150 | |
| Critical Assemblies | | ~10,000 |
| Pulsed Reactors | | ~2,000 |
| TOTAL | ~4,000 | ~12,000 |
| Amount in Traditional Scope of RERTR | 800 | |
| Previously Reduced by Conversion | 280 | 10s |
| Previously Reduced by Shutdown | 450 | ~1000s |

*Figure 1.7* Most Non-Weapons HEU Use Continues

eliminated so far. Annual use of fresh HEU has been reduced 280 kg by conversion and 450 kg by shutdown, and a few tons likely has been removed from use in lifetime cores by closure of critical facilities and pulsed reactors. But of the four tons of fresh HEU still used annually, only 800 kg – one-fifth – falls within the traditional scope of the RERTR program and thereby has been targeted for phase-out. Past phase-out efforts also have ignored the tons of HEU in the lifetime cores of critical assemblies and pulsed reactors. Although RERTR has made admirable progress, such statistics highlight the insufficiency of the program's traditional scope and actions to date. That is what prompted our research.

## Scope and Methods of the Investigation

This book investigates the prospects of a global phase-out of non-weapons HEU use. It does so by exploring past successes, continuing progress, and ongoing challenges. Our research team benefited from briefings by five experts: Parrish Staples, director of the U.S. National Nuclear Security Administration's Office of European and African Threat Reduction; Armando Travelli, founding manager of the RERTR program; Frank von Hippel, Professor at Princeton University and co-chair of the International Panel on Fissile Materials; Edwin Lyman, Senior Staff Scientist for Global Security at the Union of Concerned Scientists; and Steven Biegalski, director of the Nuclear Engineering Teaching Laboratory at University of Texas at Austin.

The researchers comprised 13 graduate students from the University of Texas at Austin – including four Ph.D. students in nuclear engineering, and nine M.A. students in public affairs – supervised by this author. They conducted field interviews in 11 countries: Argentina, Belgium, Canada, China, France, Germany, Netherlands, Norway, Russia, South Africa, and the United States.

## New Findings

The subsequent chapters provide the state of knowledge on the above topics. This section highlights the new information in those chapters that was obtained via primary research. It follows the same organization as the subsequent chapters.

### *Argentina: Medical Isotope Production*

Argentina successfully converted in 2002 to LEU targets for production of medical isotopes, which increased its production costs by 5 percent. In our investigation, the operator revealed that it converted because the United States made clear that under the 1992 U.S. Energy Policy Act, Argentina would receive no further HEU. The United States also offered at least three incentives: arranging to take back spent HEU target material from Argentina; supplying an equivalent amount of U-235 in 19.9%-enriched LEU to fabricate new targets; and helping to design an LEU target that had the same U-235 content as the previous HEU version by increasing the uranium density, thickening the fuel meat, and thinning the cladding. The operator managed to convert without interrupting production by simultaneously irradiating HEU and LEU targets during four weekly cycles over the course of a month, and then processing both types in the same line of hot cells using two different filters. After the LEU process was proved successful, the operator switched to exclusively LEU targets.

This smooth conversion was facilitated by surplus irradiation capacity in the reactor and surplus processing capacity in the hot cells, both resulting in part from the operator's relatively low output at the time. The operator subsequently has increased production by using more targets and increasing the power of its reactor. It may further boost production by switching to a foil LEU target. This case illustrates the importance of restricting HEU supply to encourage conversion to LEU. It also demonstrates that when a medical isotope producer pursues conversion in good faith, it may be able to avoid substantial cost increases, interruptions in production, or decreases in output.

### *South Africa: Reactor Fuel and Medical Isotope Production*

South Africa completed conversion of its research reactor to LEU fuel in 2009, then started using LEU targets in 2010 for some of its large-scale production of medical isotopes. It intends to complete the conversion to exclusively LEU targets as soon as it receives regulatory approval in all recipient countries for distribution of the Mo-99 produced with LEU targets, possibly by 2013. The conversion of isotope production has been facilitated by the fact that the original targets used only 45%-enriched HEU. A top South African official also told the press that the decision to convert was motivated partly by an expectation that the United States would erect "some kind of tariff barrier which would make it not as attractive to import molybdenum ... derived from HEU." The resulting conversion to LEU targets has increased current production costs by "at least 10 percent." But these costs are expected to decrease, as South Africa optimizes its production process and switches from importing French LEU targets to producing them domestically, under license from France's CERCA. Indeed, South Africa aims eventually to export LEU targets for irradiation elsewhere. The ongoing completion and optimization of conversion is being assisted by a $25 million grant from the U.S. GTRI.

South African officials note that the profitability of their business is threatened much less by the marginal cost increase from conversion than by a recent 30 percent decrease in the world market price for Mo-99. That price drop was an unforeseen long-term consequence of the temporary global supply interruption of 2009–2010, which compelled doctors to optimize their utilization procedures, thereby reducing global demand for Mo-99. This case demonstrates that prospective U.S. import preferences for non-HEU produced medical isotopes could encourage conversion by foreign producers. It also shows that the production cost increases resulting from conversion, even for a large-scale producer, may be relatively marginal, especially compared to other market factors.

### *USA and Europe: High-Power Research Reactors*

Progress continues on the original RERTR objective to convert remaining, HEU-fueled research reactors to LEU fuel. Of the nine high-power research reactors (HPRRs) that still require HEU in the United States and Europe, all but two have committed to convert and are expected to do so this decade. All five U.S. HPRRs are slated to convert to 19.75%-enriched, monolithic U-10Mo (i.e., 10 percent molybdenum) fuel of uranium density at least 15 g/cc. The plan is to convert one per year starting in 2017, in the following order: MITR, NBSR, MURR, HFIR, ATR.[11] That sequence arises because the last two reactors have more complicated fuel geometry, and the ATR is used to test the LEU fuel for all the conversions. The U.S. fuel development program has successfully irradiated full-size fuel "plates" of the monolithic LEU, so the next step is to fabricate fuel "elements" for irradiation testing.

In Europe, fuel development by the LEONIDAS program focuses on 19.7%-enriched, U-7Mo dispersion fuel of uranium density 8 g/cc in an Al-Si matrix. The program has completed lab-scale production of LEU fuel "plates" that are now being irradiated in Belgium. If successful, the next step will be lab-scale production of fuel "elements." Of the four European HPRRs, only two – BR2 in Belgium, and RHF in France – are committed to convert to LEU. The Orphée, also in France, could likewise convert to the dispersion LEU fuel, but the operator says it will await the outcome of RHF's conversion before making any decision. The final European HPRR, the FRM-II in Germany, is discussed separately below.

### *Russia: Icebreaker Ships and Floating Reactors*

Russia's small fleet of nuclear-propelled icebreaker ships lies outside the traditional scope of RERTR and still runs on HEU fuel, but prospects are good for conversion to LEU within a decade. Five ships are currently operational, another two are temporarily out of commission, and the rest are decommissioning. The fleet employs three types of reactors using fuel of enrichment that is reported varyingly between 30–90 percent. Russian propulsion-type reactors – whether used in civilian icebreakers, military submarines or surface vessels, or floating nuclear powerplants – employ a cassette fuel that can be removed and replaced with fuel of different enrichment, potentially facilitating conversion. Two of the reactor models currently used for icebreaker propulsion reactors, the KLT-40 and KLT-40M, probably could be converted to LEU, because the prototype floating nuclear powerplant uses a very similar reactor, the KLT-40S, which is fueled with LEU of 13–16 percent enrichment. Indeed, our investigation reveals that this initial reactor for the floating plant was tested in an icebreaker, which suggests that at least one icebreaker already may have run on LEU fuel.

However, Russia has no plans to convert to LEU the reactors in the current icebreakers, whose service life is being extended until at least 2018. Instead, Russia plans to use LEU fuel in the next generation of icebreaker vessels that will be powered by the new RITM-200 reactor, whose compact integrated core places the steam generator inside the reactor vessel. Russian sources say this reactor will use LEU to accord with the growing international norm for civilian facilities, even though fuel of 30–35 percent enrichment would be more efficient. Likewise, the next generation of floating powerplants is anticipated to employ a new reactor, the KLT-20, which will continue the use of LEU fuel, of 19.2 percent enrichment. Russian observance of the international norm is reportedly driven by a desire to establish an export market for both types of facilities: floating nuclear powerplants and nuclear-propelled ships.

### *Belgium and the Netherlands: Medical Isotope Production*

The world's second and third largest producers of medical isotopes, both in Europe, still use HEU targets, but both companies are likely to convert to LEU targets this decade. Their home countries, Belgium and the Netherlands, signed an agreement at the 2012 Nuclear Security Summit, pledging to eliminate HEU targets by 2015. Belgium-based IRE will convert mainly because it is running out of fresh HEU, and it does not recover remaining HEU when it processes its spent target material to remove strontium and yttrium. The United States, as a matter of policy but not current law, refuses to supply more HEU to Europe for targets to produce isotopes without a firm commitment and plan by the recipient to convert to LEU targets. IRE can convert without major new construction or interruption of production because it has two processing lines but uses only one at a time. The company does not anticipate that the licensing of any changes to its facility would hinder conversion. IRE is likely to adopt the LEU foil target developed by Argonne National Laboratory, because it fears that the alternative, silicide target could damage equipment in its hot cells. Although Argonne claims that the yield of Mo-99 per target will be unchanged by conversion to the LEU foil, IRE expects it will have to increase the number of targets, thereby necessitating additional irradiation positions in reactors, redesigned transportation containers, and possibly other expenses.

Covidien, based in the Netherlands, plans to convert to LEU targets mainly because the forthcoming Pallas reactor, which will become its main irradiation source, will not be licensed for the presence of HEU. Conversion would require either the construction of a new processing line, or some interruption in production, because the company currently uses both of its processing lines simultaneously. Delays in construction of the Pallas reactor have postponed its expected start-up from 2016 to the 2020 timeframe, which is now viewed as the ultimate deadline for conversion. Covidien also is reportedly constrained by a limited HEU supply, which may have facilitated the Dutch government's commitment to convert even earlier.

These cases demonstrate that one key to encouraging foreign isotope producers to convert to LEU is to restrict their supply of HEU from the United States, as will be done by the American Medical Isotopes Production Act, enacted in January 2013. They also underscore that European national governments, and the OECD's

High Level Group on the Security of Supply of Medical Radioisotopes, should do more to facilitate expeditious conversion to LEU, including perhaps incentivizing the purchase of isotopes produced without HEU. Finally, the impending conversion to LEU by these two European producers underscores that Russia is violating international norms by currently expanding its Mo-99 production using HEU targets and HEU fuel.

### *China: Reactors and Nuclear Propulsion*

China continues to eschew HEU for naval propulsion, and it is converting to LEU or shutting down the miniature nuclear reactors that it has constructed or exported, but it has no plans to phase out its two significant uses of HEU: for a fast reactor and the critical assembly that models it. Our investigation reveals that China initially chose LEU to fuel its nuclear submarines because of limited HEU production capacity, but it still uses LEU in second-generation vessels because such fuel is familiar and sufficient to achieve performance objectives. China's new submarines are far superior to their predecessors, demonstrating that the performance of nuclear propulsion reactors can be enhanced significantly without increasing enrichment. Although China does not use HEU for propulsion, it opposes inclusion of naval fuel in a Fissile Material Cutoff Treaty (FMCT) because it fears this would compel it to declare its HEU inventory, which it does not want to do. (That fear does not appear warranted, because even if the draft FMCT were expanded to ban HEU for naval propulsion, the nuclear-weapons states such as China would be required to declare only their HEU inventories in excess of weapons requirements.) In any case, China probably will not voice its opposition to including naval HEU in the FMCT, because it expects the United States and Russia to exclude that category of fissile material from the treaty.

China and the GTRI, in September 2010, signed an agreement to construct an LEU critical assembly in China to facilitate conversion of two of its domestic miniature reactors and five that it previously exported to Ghana, Iran, Nigeria, Pakistan, and Syria – each of which contains approximately 1 kg of HEU. (Sources reveal that China is shutting down its other remaining miniature reactor, at Shandong.) The total amount of HEU in these facilities is relatively small, but converting them bolsters the global norm against non-weapons HEU. By contrast, China has not discussed shutting down, converting to LEU, or reducing the enrichment of its fast reactor – which contains 240kg of 64%-enriched uranium – nor the critical assembly that models it using 41kg of 90%-enriched HEU. These two facilities eventually are planned to convert to MOX fuel. Another reactor, the PPR, is sometimes erroneously listed as HEU-fueled, but sources confirm that its fuel is 19.9%-enriched LEU. Although China's record on non-weapons HEU is admirable, its government could contribute further to a global HEU phaseout by publicizing its use of LEU in second-generation submarines, embracing the inclusion of naval fuel in an FMCT, and beginning to explore reducing the enrichment of its fast reactor and critical assembly.[12]

### *Germany: The FRM-II Reactor*

The FRM-II now annually uses four cores (down from the original five), each containing 8 kg of 93%-enriched HEU, for an annual demand of 32 kg. The reactor had been required to convert by 2010 to fuel of 50%-enrichment or lower, but because suitable fuel was not available, this deadline was extended to 2018. The reactor's current supply of fuel runs out in 2015–2016, and it is not clear where more HEU will be obtained to fuel the reactor until at least 2018, which could require 96 kg or more. Our investigation revealed an unpublished paper by the facility's operator at the Technical University-Munich that examines the feasibility of conversion. We obtained a portion of this report, indicating that it is feasible to reduce the fuel's enrichment, but not all the way down to LEU. The analysis contends that the core could be converted, presumably without changing its geometry or the reactor's power, either to 50%-enriched dispersion fuel of 8 g/cc developed by LEONIDAS (with an 8 percent reduction in neutron flux), or to 34%-enriched monolithic fuel of 15 g/cc developed by the U.S. GTRI (with a flux penalty exceeding 10 percent). However, a study at Darmstadt University affirms the U.S. Argonne National Laboratory's contention that the FRM-II could be converted to 19.9%-enriched LEU with a flux penalty of less than 10 percent, if the core's height were increased from 10cm to 12cm and the reactor's power increased from 20 MW to 22 MW. Regarding press reports of the facility planning to produce Mo-99 for medical isotopes, it was learned that the FRM-II will only irradiate targets for an outside producer, who will determine the enrichment of the targets and will process them offsite.

German experts suggest that the FRM-II is likely to undertake at most one conversion. Moreover, a reduction to 50 percent enrichment would provide only modest security benefits – increasing the critical mass by 3 to 5 times, depending on weapons design – and would tend to undermine the international norm of conversion to LEU. Thus, the path most consistent with a global HEU phase-out would be to modify the core's dimensions slightly so that it could be converted to LEU, or barring that to convert the core to 34%-enriched monolithic fuel. International leverage to compel such a substantial reduction of the enrichment may be available on both ends of the fuel cycle. That is because the reactor has an impending need for fresh fuel and lacks of a disposition method for its spent fuel, which by virtue of its high enrichment and silicide matrix presents unique back-end challenges.

### *Canada and Russia: Medical Isotope Production*

Nordion is the largest supplier of the world's main medical isotope Mo-99, which it currently produces by irradiating HEU targets at Canada's LEU-fueled NRU reactor. But that reactor will cease isotope production in 2016 and its two planned successor MAPLE reactors have been canceled. Accordingly, in September 2010, Nordion signed a contract to become the exclusive international distributor of Mo-99 produced by the Russian firm ISOTOPE, a subsidiary of Rosatom. Russia traditionally has produced Mo-99 for domestic consumption using the Karpov VVR-T reactor in Obninsk, near Moscow. ISOTOPE will instead produce Mo-99 for export by irradiating HEU targets in three HEU-fueled research reactors – the SM, RBT-6, RBT-10 – at RIAR in Dimitrovgrad.

ISOTOPE is rapidly implementing a three-phase expansion. In Phase 1, which was completed in January 2011, it increased production to 1,000 six-day Ci/wk by processing the irradiated targets in existing hot cells. In Phase 2, during 2011–2012, it has been gradually increasing production by starting two processing lines in its new IDU target processing facility, built by the German company Isotope Technology Dresden (ITD). In Phase 3, projected initially for early 2012 but now delayed for unknown reasons, it aims to produce roughly 3,000 six-day ci/week, about one-quarter of the global demand for Mo-99. Russian officials say that in Phase 3, after establishing ISOTOPE as a global supplier, they intend to convert to LEU targets by 2017. History suggests, however, that conversion to LEU would be easier, less expensive, and more likely to occur, if undertaken prior to major investment to expand production using HEU targets.

### *Russia: Research Reactors*

Although the Soviet Union implemented a program in the 1980s to reduce the enrichment of fuel in exported reactors, Russia had refused to consider conversion of its own domestic reactors until, in December 2010, it signed an agreement with the United States to explore the feasibility of converting six Russian reactors: IR-8, OR, ARGUS, IRT, IRT-T, and MIR.M1. Russian officials say the deal was facilitated by recent U.S. commitments and steps by Washington to convert all U.S. research reactors to LEU, demonstrating that high-quality nuclear research does not require HEU. The bilateral deal is potentially very important because Russia uses more HEU in research reactors than any other country. Technical feasibility is not an obstacle to conversion, Russian officials told us, because high-density LEU-Mo dispersion fuel, currently under development in Russia, is envisioned as sufficient to convert all Russian reactors that currently use HEU except the SM-3 at Dmitrovgrad. Russian officials project successful development of the LEU-Mo fuel within 2–3 years, but U.S. officials warn that licensing, production, and implementation of this fuel would require 5–10 years. Studies on four of the six Russian reactors were completed by June 2012, finding conversion to LEU fuel to be feasible, and the other studies are ongoing. As a result, Russia says it will convert the first one or two of the reactors in 2014.[13]

The main obstacles to conversion are economic. Russian officials confirm that salaries at nuclear facilities are currently tied to the level of fuel enrichment, so scientists can be expected to oppose conversion to LEU that would cut their income by up to half. Moreover, the proposed LEU fuel is projected to cost significantly more, and neither Rosatom nor the facility operators are eager to pay for it. In practice, it is unclear whether Rosatom can impose conversion on some facilities that fund their own operations through commercial activity, such as production of medical isotopes. Likewise, the new PIK reactor, fueled by

90%-enriched HEU and due to start operations in 2014, has an autonomous management body that could potentially block conversion. Russia's only possible economic incentive for conversion is to prove the feasibility of its LEU-Mo fuels so that they could be marketed for export. But in light of the other economic obstacles, it may be necessary for the United States and/or the international community to provide financial support, or other incentives,[14] to persuade Russia to convert more of its research reactors to LEU.

### *Russia: Critical Assemblies and Pulsed Reactors*

Our investigation provides an updated database, including location where available, of Russia's critical assemblies and pulsed reactors. These facilities raise significant risks because of their large quantities of very lightly irradiated HEU – for example, the Sarov critical assembly by itself contains 833 kg of 90%-enriched uranium. Russia legitimately may need a few HEU-fueled pulsed reactors and critical assemblies, but other countries have replaced such facilities almost entirely with computer simulations. Indeed, Europe has no more than four remaining critical assemblies fueled with 90%-enriched uranium, but Russia still has roughly 30 of them. Russian officials claim that they conduct a cost-benefit analysis for each facility, weighing the risk and expected cost of an HEU theft versus the cost of conversion or decommissioning.

For critical assemblies, as with powered research reactors, the obstacles to conversion or decommissioning are mainly economic: the fear of decreased salaries, increased fuel costs for conversion, and unfunded expenses for decommissioning. In addition, Rosatom may lack the political leverage and/or funding to demand conversion, especially at facilities that engage in important research or have commercial contracts. Accordingly, some Russian operators express hope for U.S. financial assistance to enable conversion or closure of superfluous critical assemblies in the civilian sector. Russian officials refuse to discuss the majority of pulsed reactors that are in the defense sector. They are willing to contemplate conversion of the few pulsed reactors in the civilian sector, such as the Hydra at Kurchatov, but say that some others of these would be difficult to convert because they already use higher than normal density HEU fuel, so that substitute LEU fuel might lack sufficient concentration of U-235. The United States could encourage conversion or decommissioning of Russia's critical facilities by providing computer simulation codes and financial assistance.

### *USA and France: Naval Propulsion*

Although military facilities are relatively well protected against theft of fissile material, the U.S. Navy's continued use of HEU in propulsion reactors tends to legitimize such fuel for other high-performance activities and thereby undermines a global phase-out. Our investigation reveals that France, by contrast, has now converted all of its naval nuclear propulsion to LEU fuel, contrary to some claims in the literature. France made this decision because its HEU production was costly and LEU fuel was sufficient to meet its strategic objectives. Conversion was accomplished by using high-density, 10.25 g/cc "caramel" LEU fuel, in a compact, integral core design that places the steam generator inside the reactor vessel. France continues to improve reactor performance even as it further reduces enrichment. The reactor slated for its next generation of submarines will have its core life extended by 43 percent, from 7 years to 10 years, while the fuel's enrichment will be reduced from the current 7 percent to a commercial level under 5 percent.

The U.S. Navy cites two main grounds for opposing conversion – economics and performance – because LEU reactors would have larger size, less power, and/or shorter core life. In economic terms, our investigation reveals that the United States has never seriously studied the full costs and savings (for example, from reduced security and fabrication costs of LEU fuel) of an optimized conversion, based on designing a next generation of vessels rather than retrofitting existing ones. On performance, LEU does have the disadvantage of a negative U-238 absorption reactivity factor, so that unlike an HEU reactor in which power can be controlled simply by changing the amount of reactor-produced steam that enters the vessel's turbines, with an LEU reactor the operator must also manipulate control rods, adding time and a potential source of error. However, based on our discussions with a senior manager of naval programs and an academic expert, this hurdle is likely surmountable. The most realistic path forward, consistent with efforts toward a global HEU phase-out, is not to convert the existing U.S. naval fleet but to start immediately to explore the trade-offs and designs that would be necessary for the next generation of U.S. naval vessels to utilize LEU fuel.

### *Russia: Naval Propulsion*

Russia's naval reactors use HEU fuel of medium enrichment, and could technically be converted to LEU, but instead may have their fuel enrichment increased to extend core life. This case is important because Russia's navy annually demands an estimated 570 kg of fresh HEU, roughly one seventh of global non-weapons HEU use. Each submarine core contains an average 115 kg of HEU, of 21–45 percent average enrichment (though certain zones may be enriched up to 70 percent),[15] lasting about eight years. Russia originally chose this medium enrichment of HEU to enhance the reliability and safety of its submarine reactors, while the reactor in its one missile cruiser uses HEU of higher, 55–90 percent enrichment. Russia's next generation of naval vessels is likely to be powered by the same RITM-200 reactor with compact integrated core that is expected for civilian icebreaker ships. Although this reactor is slated to use LEU fuel in the icebreakers, it may use HEU in submarines because 30–35 percent enrichment is reportedly more efficient, and the cassette fuel makes this swap feasible. (One source suggests that a precursor of this reactor already has been deployed in recent *Borei* class submarines.) More generally, the Russian Navy reportedly plans to *increase* the enrichment of fuel in its propulsion reactors to extend core life to match the longer maintenance cycle of its vessels.

It would be technically possible, however, for Russia to convert its naval vessels to LEU fuel using the new RITM-200 reactor. Sources in Russia confirm that they have sufficient LEU fuel fabrication capacity and no backlog of spent cores or reactors awaiting reprocessing or decommissioning that would hinder conversion. Russian officials, however, express no intention to convert naval propulsion reactors to LEU, noting that upgrades in MPC&A (material protection, control and accounting) have improved security and that the U.S. Navy has not declared any plan to convert its propulsion reactors. The only possible internal incentive for Russia to convert its naval propulsion to LEU would be to capture economies of scale from using the same fuel anticipated for icebreaker ships and possibly floating powerplants. The United States could encourage conversion of Russia's nuclear navy to LEU fuel by committing to convert the U.S. Navy fleet.

### *Space Reactors*

Space reactors pose potentially the greatest obstacle to a global phase-out of non-weapons HEU, because of the exigency in space missions to minimize weight and radioactivity, as HEU fuel does. Indeed, some critics would argue that space reactors should be excluded from an HEU phase-out, because after launch the reactors do not themselves increase risks of nuclear proliferation and nuclear terrorism. However, HEU-fueled space reactors do tend to legitimize such fuel for other non-weapons applications, and their ground-based HEU infrastructure does entail risks, which could spread as additional countries create space programs.

Although there have been no launches in recent years, from 1967 to 1988, the Soviet Union launched into space 35 nuclear reactors, all fueled with HEU. The reactors were intended for either electric or thermal propulsion, typically to power satellites for tracking U.S. naval vessels. The United States launched one reactor into space, in 1965, also HEU-fueled. (Space reactors should not be confused with radioisotope thermoelectric generators, which produce smaller amounts of energy for space missions from the radioactive decay of plutonium-238 or other elements.) No country is known currently to be planning a launch, but several are contemplating future space reactors, still based on HEU fuel. In a 2009 paper, Japanese scientists explored an LEU-fueled reactor design for a notional interstellar probe, but no country has conducted a serious assessment of the penalties to reactor weight or mission capability (or any possible benefits) from converting to LEU.

In light of the current pause in space-reactor launches, countries contemplating future ones should utilize this window of opportunity to conduct feasibility assessments of potential conversion, before that is precluded by serious planning of HEU-fueled missions. In the United States, the White House Office of Science and Technology Policy could ask the Department of Energy to engage the RERTR program for such a study, building upon a preliminary assessment conducted in 1994 by the DOE's Office of Nuclear Energy. The international community also should correct a drafting error in the 1992 UN General Assembly resolution on space reactors that aimed to ban plutonium fuel, but did so imprecisely by mandating that such reactors must use HEU fuel. An amendment should clarify that LEU is not only acceptable but preferable if feasible.

## How to Promote an HEU Phase-Out

The above case studies give rise to a series of prescriptions (summarized in Figure 1.8) to promote a global phase-out of HEU commerce for non-weapons purposes, which could also facilitate a worldwide ban on production of HEU:

### *Preferential Procurement*

All countries – but especially the large consumers including the United States that can shape markets – should establish a preference for the purchase and use of Mo-99, and resulting medical pharmaceuticals, produced without HEU. This could be pursued either by taxing the HEU-based versions or subsidizing the non-HEU-based versions (the U.S. government is pursuing the latter strategy),[16] or by banning the import or use of the HEU-based versions as soon as a sufficient supply is available produced without HEU.

### *Conditional HEU Supply*

All countries – but especially Russia and the United States, which are the two traditional exporters of HEU – should condition any further export on the recipient pledging to phase-out its reliance on HEU as soon as possible. In the United States, this approach is embodied in the Schumer Amendment to the 1992 Energy Policy Act, which was partially relaxed by the Burr Amendment to the 2005 Energy Policy Act for Mo-99 production. As a matter of policy, the U.S. government and Nuclear Regulatory Commission should apply the principle of the 1992 law to any HEU export, including for Mo-99 production, and Russia should adopt a similar policy.

This approach could prove immediately effective in two instances. First, Germany's FRM-II reactor is anticipated to exhaust its supply of HEU fuel by 2016, at which time the ultra-high density LEU fuel is expected to be qualified. If both Washington and Moscow imposed conditionality on HEU supply, the FRM-II's operator would have little choice but to substantially reduce the enrichment of

- Preferential Procurement
- Conditional HEU Supply
- Conditional Take-Back of Spent HEU
- Publicize Outliers
- Subsidize Conversion and Shutdown
- Technical Assistance
- Feasibility Studies
- Conditional Commitment on Naval Reactors
- Expand FMCT

*Figure 1.8* Policy Options to Promote Global HEU Phase-Out

its fuel, ideally to LEU. Second, Europe's two large producers of Mo-99, in the Netherlands and Belgium, have nearly exhausted their supply of HEU for targets, and reportedly there is no additional supply in Europe. If Russia and the United States imposed conditionality, both producers would be compelled to convert by approximately 2015, as their national governments have committed to do, while in the interim receiving small quantities of HEU to avoid interruption in production.

### *Conditional Take-Back of Spent HEU*

As with the front end of the fuel cycle, its back end provides leverage to encourage the phase-out of HEU commerce. Russia and the United States should accept the return of spent HEU from facilities that still use such fuel only if the operator pledges to phase out reliance on HEU as soon as possible. Germany's FRM-II is a prime example again. The reactor's HEU spent fuel contains silicide, making it unattractive to France's AREVA for reprocessing and thereby leaving the operator without a permanent back-end solution, which increases its costs and could endanger its operating license. If the United States and/or Russia offered to accept the spent fuel, but only on condition that HEU be phased out, this too could compel the operator to significantly reduce the enrichment of its fuel, ideally to LEU.

### *Publicize Outliers*

In cases where operators have resisted phasing out HEU, and no leverage is provided by a need for fuel supply or take back, the best approach is to publicize the facility's non-compliance with the growing international norm to minimize HEU commerce – an approach known as "naming and shaming." For example, in the realm of European and U.S. high-power research reactors, only two of nine facilities are resisting conversion to LEU, and one of them, Germany's FRM-II, is susceptible to leverage on its fuel cycle. That leaves only France's Orphée reactor, which probably could convert to ultra-high-density LEU fuel after 2016 but has refused to commit to this, or even to explore its feasibility. The operator apparently has a sufficiently assured supply of fresh HEU fuel and a back-end solution for its spent fuel, presumably reprocessing by AREVA. Therefore, the best remaining approach to encourage conversion is to publicize that this facility is out of step with the prevailing international norm, undermining international efforts to minimize HEU commerce and thereby potentially increasing global risks of nuclear proliferation and nuclear terrorism. French and foreign NGOs, journalists, and government officials should ask the operator, repeatedly and in public forums, to explain its resistance, including any ostensible obstacles to conversion.

### *Subsidize Conversion and Shutdown of Russian Facilities*

In Russia, our research identified economic costs as a primary obstacle to the elimination of HEU use – via conversion to LEU fuel or shutdown – especially for the following types of facilities: critical assemblies, pulsed reactors, and existing propulsion reactors in icebreaker ships. Potential costs arise from various factors: shutdown causes layoffs and decommissioning expenses; the mere consideration of conversion may prompt shutdown and its economic consequences; actual conversion to LEU may trigger a reduction in staff salaries; high-density LEU fuel may cost more than traditional HEU fuel; and converting a partially irradiated core prior to full burn-up necessitates the purchase of a new core earlier than anticipated. For all these reasons, the provision of subsidies to operators could facilitate the conversion to LEU or shutdown of such facilities.

The security benefits of phasing out HEU use in Russia would accrue beyond that country as well, given the international nature of terrorist and criminal networks and the easy transportability of the small quantities necessary for weapons. Accordingly, it is appropriate for other countries, with sufficient means and interest, to consider helping Russia financially to subsidize the cost of eliminating HEU use at these facilities. Russia also should modify policies that pose disincentives to HEU phase-out. For example, to promote conversion of facilities to LEU, Russia should cease tying salaries to the enrichment level of fuel. To promote the possibility of shutdowns, Russia should guarantee continued salaries or equivalent pensions to workers at affected facilities.

### *Technical Assistance to Russia*

Another obstacle to the shutdown of Russia's critical assemblies and pulsed reactors is that, unlike the rest of the industrialized world that uses computer codes to design reactors and model the effects of nuclear weapons,[17] Russia's engineers and scientists depend on experiments at these facilities. Certainly, computer codes for the modeling of neutronics and thermal dynamics do need to be benchmarked by experiments, but in most of the world this is accomplished at very few facilities, not dozens. Indeed, the IAEA reportedly confirmed as early as 2005 that

> computer simulation of reactor cores has become good enough that, for standard types of reactors, they only need to be tested against benchmark experiments, and enough archived criticality experiments are available to provide benchmark tests in such cases. It should therefore be possible to retire and decommission many critical facilities.[18]

Russia's divergence in this regard may result from two factors: a lack of sophisticated computer codes; and the culture in its nuclear community of distrusting computer codes in favor of actual experiments. Other countries could help by sharing computer codes with Russia and working to develop them and implement their use. The United States and others might hesitate to do so because they view such codes as vital to national security, even restricting their export as munitions. But these concerns should be balanced against the security risk posed by the ongoing use in Russia of tons of weapons-grade uranium at dozens of ostensibly non-weapons facilities.

### *Expand Scope of Feasibility Studies*

Just as feasibility studies facilitated the RERTR program's original mission of converting research reactor fuel and Mo-99 production targets to LEU, now it is necessary to expand the scope of such investigations. Two obvious categories for study are Russian critical assemblies and civilian pulsed reactors (which could lay the groundwork for future assessments of military pulsed reactors that are more politically sensitive). Naval propulsion conversion studies also should commence immediately, even if U.S. and Russian procurement of next-generation submarines is years away, because any reactor modification could affect design of the larger vessel. Likewise, for space reactors, the current pause in launches provides a window of opportunity to explore trade-offs entailed by conversion to LEU fuel, and to modify the 1992 UN resolution to make clear that such fuel is an acceptable – indeed, preferable – alternative to HEU.

### *Conditional Commitment on Naval Reactors*

Although France and China demonstrate the feasibility of LEU fuel for propulsion of military submarines, the four states (U.S., Russia, UK, India) that now use HEU fuel in naval reactors may hesitate to convert to LEU even for their next generation of vessels due to fear of decreased performance. This concern could be addressed by a *conditional* conversion commitment, which could take three forms.

First, a state could declare its *unilateral* commitment to convert its next generation of naval propulsion to LEU on condition that it proves technically feasible. This would be analogous to the commitment that many states and operators have made previously with regard to converting research reactors and Mo-99 production, and that is required by the 1992 Energy Policy Act as a condition of HEU exports. As a domestic example, in 1986, the U.S. Nuclear Regulatory Commission ordered all licensed research reactors to convert when feasible, even though such feasibility was not proved for high-power reactors until 25 years later.

If the two countries with the largest nuclear navies – the United States and Russia – resist such a unilateral commitment, an alternative would be a conditional *bilateral* reciprocal commitment. For example, the United States could declare that it will convert its next generation of naval propulsion reactors if Russia agrees to do the same. But it might be even more advantageous for Russia to seize the initiative and challenge the United States to reciprocate. That is because Russia's naval fuel is enriched to an average of only about 35 percent, compared to 97 percent in U.S. propulsion reactors, so any cost and performance penalties for reducing enrichment below 20 percent should be more onerous for the United States. Thus, Russia might obtain both international goodwill and some relative gain by declaring its commitment to convert its naval reactors to LEU on condition that the United States reciprocates.

Eventually, it would be ideal to forge a *multilateral* treaty barring HEU for naval propulsion. This would entail both the UK and India also converting their next generation of submarine reactors. Such an international approach would provide greater incentive for any additional states that acquire nuclear navies to use LEU fuel, given the enhanced confidence that no other state would gain a competitive advantage by using HEU. However, in light of the likely substantial delay before such a treaty could be adopted, the phase-out of HEU for naval propulsion should start via the unilateral or bilateral approaches discussed above.

### *Expand FMCT*

As noted at the outset, any continued use of HEU for non-weapons purposes would hinder adoption of the proposed Fissile Material Cut-Off Treaty, because states could produce HEU for ostensibly non-weapons purposes and then divert it to weapons. The goal, therefore, should be to broaden the FMCT to bar all production of HEU for any purpose. Realistically, given that certain states have large stockpiles of HEU for non-weapons uses, and other states would not want to face a competitive disadvantage in this realm, the treaty probably would also need to ban non-weapons uses of HEU. Thus, a broadened FMCT could ban production of HEU, use of HEU for non-weapons purposes, and production of plutonium for weapons. That would not address two other concerns – nuclear disarmament and plutonium production for non-weapons purposes – but those more contentious issues should not be permitted to hinder progress on the rest, where there is greater consensus that could enable significant gains in international security.

## Conclusion

More than three decades of international efforts have succeeded at phasing out hundreds of kilograms (about a dozen bombs' worth) of annual non-weapons HEU use. These initiatives understandably targeted facilities on college campuses, at private manufacturers, and in developing countries that were among the most vulnerable to theft or diversion for nuclear weapons by states or terrorists. Progress continues, so that by the end of this decade, HEU is slated to be virtually eliminated from research reactors and medical-isotope production worldwide, except in Russia.

Unfortunately, such past and anticipated progress is insufficient to address the magnitude of the challenge. Even if current expectations are met for conversion and shutdown, the vast majority of non-weapons HEU use will continue at facilities outside the traditional scope of these efforts – in critical assemblies, pulsed reactors, naval propulsion, and anticipated space reactors. Moreover, as terrorists and proliferant states learn that less HEU is vulnerable at research reactors and isotope production plants, they are likely to turn their attention to the other facilities, which in many cases contain even larger amounts of HEU.

To address this looming threat, the international community should consider expanding past efforts into a total, global phase-out of HEU for non-weapons purposes. The preceding section offered nine concrete steps to promote this goal. This includes expanding the Fissile Materials Cut-Off treaty to incorporate a worldwide ban on production of HEU, which would simultaneously promote three vital goals: arms control, nonproliferation, and anti-terrorism.

Such grand objectives may seem unrealistic. But if nothing else, the chapters in this book are testament that seemingly insurmountable technical, economic, and political obstacles to phasing out HEU commerce have been, and continue to be, overcome repeatedly – by scientific ingenuity, financial support, and far-sighted statesmanship. The only real obstacle to a global phase-out of HEU, therefore, may be the adequacy of political will. Our sincere hope is that this book offers both insight into past achievements and inspiration for the future efforts that will be necessary to address threats of nuclear proliferation and nuclear terrorism.

## Notes

1 The author thanks Dr. Frank von Hippel for helpful comments on a preceding draft.

2 Luis Alvarez, *Adventures of a Physicist* (Basic Books, 1987), p. 125

3 Joseph L. Sapir, Russell Kidman, and R. W. Brewer, "235U(94%) Spheres Surrounded by Natural Uranium Reflectors," LA-UR-98-1664, Los Alamos National Laboratory, 1998, http://lib-www.lanl.gov/la-pubs/00418688.pdf (accessed on 27 September 2012).

4 *IAEA Safeguards Glossary*, 2001 Edition, http://www-pub.iaea.org/MTCD/publications/PDF/nvs-3-cd/PDF/NVS3_prn.pdf (accessed on 27 September 2012). Perhaps coincidentally, this amount is roughly the critical mass of a sphere of 94%-enriched HEU surrounded by a 1.76-inch reflector of natural uranium.

5 National Research Council, *Medical Isotope Production Without Highly Enriched Uranium*, January 2009, http://www.nap.edu/catalog/12569.html (accessed on 27 September 2012).

6 See, Ole Reistad and Styrkaar Hustveit, "HEU Fuel Cycle Inventories and Progress on Global Minimization," *The Nonproliferation Review* 15, 2 (July 2008): 265–287, and appendices.

7 Frank von Hippel and his co-authors have highlighted this shortcoming. See, for example, Frank von Hippel, "A Comprehensive Approach to Elimination of Highly-Enriched-Uranium From All Nuclear-Reactor Fuel Cycles," *Science & Global Security* 12, 3 (2004): 137–164; Alexander Glaser and Frank N. von Hippel, "Global Cleanout: Reducing the Threat of HEU-Fueled Nuclear Terrorism," *Arms Control Today* (January/February 2006): 18–23; Alex Glaser and Frank von Hippel, "Thwarting Nuclear Terrorism," *Scientific American* (February 2006): 56–63; Chunyan Ma and Frank von Hippel, "Ending the Production of Highly Enriched Uranium for Naval Reactors," *The Nonproliferation Review* (Spring 2001): 86–101; Frank von Hippel, "The Effort to Eliminate HEU Use in Reactor Fuel," presentation to Physics Student Group, Jadwin 111, 11 April 2006, in author's possession.

8 When a critical facility is a mockup of a research reactor, the critical facility is often converted to LEU at the same time as the reactor. Frank von Hippel, personal communication, 24 July 2011.

9 Andrew Bieniawski, "An Overview of the NNSA Global Threat Reduction Program's Goals for HEU Minimization," presentation at the RERTR Annual Meeting, Beijing, China, 2 November 2009.

10 "Breakout of GTRI Reactor Conversion Expansion," NNSA, U.S. Department of Energy, 3 November 2009. "66 Reactors Added to Scope," NNSA, U.S. Department of Energy, 2011. Both documents in author's possession. The latter document states: "The scope was originally expanded by 71 reactors, but since that time 5 additional reactors have been verified as shutdown [SIC] and are now included in the list of 76

reactors ‘converted or shutdown.’” See also, email to author from Jeff Chamberlin, Deputy Director, NNSA/GTRI Office of European and African Threat Reduction, NA-212, 17 February 2012, in author’s possession. See also, Jeff Chamberlin, “Russia – United States Research Reactor Conversion Cooperation,” 15 October 2012, reporting continued progress: “To date, the program has converted or verified the shutdown of 83 reactors from a scope of 200.”

11 Frank Munger, “Eight more years of HEU fuel at HFIR,” http://blogs.knoxnews.com/munger/2012/03/eight-more-years-of-heu-fuel-a.html (accesssed 27 March 2012).

12 A large fast-reactor core could still operate with a reduction to 20–30 percent enrichment of its fuel. Smaller cores require higher enrichment due to neutron leakage. Frank von Hippel, personal communication, 24 July 2011.

13 “Russia to Convert Two Reactors to LEU in 2014,” RIA Novosti, 26 June 2012, http://en.rian.ru/russia/20120626/174248257.html (accessed 1 July 2012).

14 Alexander Glaser has suggested that conversion to LEU fuel could be encouraged by offering also to enhance neutron beam tubes to yield a net increase in neutron intensity for researchers. Frank von Hippel, personal communication, 29 July 2011.

15 A. Nikitin, Skype interview by Yaroslav Primachenko, Oslo, Norway, 17 February 2011. Nikitin is the first expert to report that some zones in the cores of Russian naval reactors are enriched above 45 percent.

16 Ian MacLeod, “U.S. anti-proliferation move could harm Nordion’s medical isotopes business,” *The Ottawa Citizen*, 9 August 2012, http://www.ottawacitizen.com/health/anti+proliferation+move+could+harm+Nordion+medical+isotopes+business/7067351/story.html (accessed on 27 September 2012).

17 For a discussion of recent work in this area, see http://www.itaps.org/applications/gnep.html (accessed on 27 September 2012).

18 Unofficial summary of findings of the IAEA’s February 2005 consultation on the “Future Use of Critical and Subcritical Assemblies,” as reported by one of its participants. See, Frank von Hippel, “Future Needs for HEU-Fueled Critical Assemblies,” paper presented to the 27th Annual International Meeting on Reduced Enrichment for Research and Test Reactors, Boston, MA, 6–10 November 2005, http://www.rertr.anl.gov/RERTR27/PDF/S9-3_vonHippel.pdf (accessed on 27 September 2012).

# Part II

# Success Stories

# 2 Argentina

## Medical Isotope Production

*Jared Berenter*

The vast majority of nuclear-medicine procedures require technecium-99m (Tc-99m), a decay product of the medical isotope molybdenum-99 (Mo-99). Mo-99 thus provides an indispensable function in the diagnosis and treatment of cancer, heart disease, and other life-threatening diseases. However, it also may create a proliferation risk. Ninety-five percent of the global Mo-99 supply is produced using targets of highly enriched uranium (HEU), the same material used to make nuclear weapons. Medical isotope producers use at least 50 kg of HEU each year for this purpose, yet the process consumes only 5 percent of the HEU's fissile U-235 isotope, leaving large quantities of vulnerable weapons-grade HEU to accumulate as "waste" as well.[1] The proliferation and terrorism concerns from fresh and spent HEU have caused many in the policymaking, medical, and nonproliferation fields to seek alternative forms of medical isotope production, including techniques that instead use low enriched uranium (LEU), which is not suitable for weapons.

A potentially competing consideration is the reliability of the global medical isotope supply. According to an International Atomic Energy Agency (IAEA) staff report, issued in January 2011, "technetium-99m is used nearly every second in some 30 million medical imaging procedures a year, making it the most extensively used diagnostic medical isotope."[2] In fact, the IAEA projects that Tc-99m consumption will continue to grow at an annual rate of three to five percent. Because of the rapid rate of decay of Mo-99, which has a half-life of about 66 hours, and even more rapid rate of decay of Tc-99m, which has a half-life of only 6 hours, isotope producers must produce year-round and minimize interruptions. While the necessary technology exists for the world's four major producers of Mo-99 – in Canada, the Netherlands, Belgium, and South Africa – to convert to using only proliferation-resistant LEU targets, they have cited a number of deterrents to conversion, including costs, interruption of production, and the risk of degrading isotope quality.[3]

These two policy goals – phasing out the use of HEU targets in medical isotope production, and ensuring reliability of isotope supply – can appear to run counter to one another. This helps explain the schizophrenic nature of U.S. policy over time, alternately favoring one of these objectives over the other. The underlying facts are that United States consumes half the world's Mo-99, while producing none; it has supplied the majority of HEU for global medical isotope production, but is also the leading advocate of eliminating such HEU use. The first U.S. statutory effort to encourage conversion to LEU targets came in 1992, as an amendment by Rep. Charles Schumer to the 1992 Energy Policy Act. That law made the export of U.S.-origin HEU conditional upon commitments from foreign users to convert to LEU.[4] The George W. Bush Administration reinforced this stance in 2003, requesting funding to facilitate conversion of worldwide Mo-99 production to use of LEU targets, under the Reduced Enrichment for Research and Test Reactors (RERTR) program.[5]

But such efforts to phase-out HEU soon confronted new hurdles from an amendment to the Energy Policy Act of 2005, sponsored by Rep. Richard Burr. The Burr Amendment allowed U.S. HEU exports to medical isotope producers in Europe and Canada without requiring a commitment to convert to LEU. [6] As observed by Loukianova and Hansell, "the Burr Amendment made it easier for Canadian and Western European Mo-99 manufacturers to continue using HEU to produce medical isotopes without cooperating fully with programs like RERTR to work toward conversion."[7]

In this light, it is significant that in 2002 Argentina became the first country to modify its medical isotope production process from HEU to LEU targets. Given Argentina's success, this case may offer lessons for overcoming the economic and political obstacles to conversion by other isotope producers, as well as for understanding the role of U.S. foreign policy in facilitating conversion. Ultimately, Argentina's example may offer insights into strategies for reconciling nonproliferation objectives with isotope supply concerns.

Some aspects of the Argentine case have previously been documented, including Argentina's role as a global leader in high-density LEU fuel fabrication. But this chapter also assesses *why* Argentina decided to convert, in light of the costs of conversion and its potential impact on the reliability of medical isotope supply. The chapter draws upon not only the existing literature, but also new field interviews with officials at Argentina's National Atomic Energy Commission (CNEA) and Nuclear Regulatory Authority (ARN). It details the conversion process in Argentina, paying keen attention to incentives, expectations, outcomes, and prospects for scaling up LEU-based production. Ultimately, I find that Argentina chose to convert because the United States discontinued HEU supply under the Schumer Amendment. This leads me to project that, with proper enforcement, the United States could use its leverage as an HEU supplier to induce larger-scale producers to convert as well.

The remainder of this chapter is divided into four parts. First, I review previous literature on the challenges of conversion to LEU targets and the efforts by the Argentines to complete the process. This section pays particular attention to the technical aspects of conversion, which the literature covers more extensively than economic or political considerations. But this section also begs the question of why Argentina converted, which has been largely ignored in previous analyses. Second, I provide a brief explanation of my methodology, including the process for selecting interviewees. Third, I detail my findings. Finally, I offer a series of policy recommendations.

## Background and Literature Review

Argentina is a small-scale producer of medical isotopes for domestic consumption and export to other Latin American countries. Isotopes are produced by CNEA, the Argentine government's nuclear agency, in the RA-3 reactor at the Ezeiza Atomic Center.[8] Until 2002, Argentina irradiated HEU targets to produce medical isotopes. In 2000, however, the country began a program to develop an LEU-based target that would work in its RA-3 reactor. The end result was a high-density LEU-aluminum dispersion target.

Information about the technical aspects of Argentina's conversion and continuing developments in LEU target fabrication is abundant. CNEA officials present annually at RERTR conferences, where they provide updates on Argentina's nuclear activities, including various aspects of LEU target fabrication and Mo-99 production using LEU targets. CNEA officials present similar progress reports at annual conferences held by the European Nuclear Society (ENS).[9]

CNEA's presentation at the 2009 RERTR meeting provides a helpful synopsis of Argentina's 2002 conversion to LEU-based isotope production. The paper highlights the following facts: "CNEA converted to LEU-based production using the same set of hot cells that were being used for HEU-based production;" no interruption in Mo-99 production occurred; the "new LEU technology satisfies the most stringent requirements of quality for its use in nuclear medicine applications;" and "overall costs for LEU-based production compared to HEU-based production increased by 5 percent." The report concludes that "no technical, quality, or financial reasons make disadvantageous [sic] to change from HEU to LEU radiochemical technology for Mo-99 and other radioisotopes production."[10]

A more comprehensive description of Argentina's conversion is found in the 2009 report by the U.S. National Academy of Sciences (NAS), *Medical Isotope Production Without Highly Enriched Uranium.* The NAS report details the uranium-aluminum dispersion targets used by CNEA. By almost doubling the density of uranium in the target meat, and thickening the meat by reducing the cladding thickness, CNEA squeezes almost half as much U-235 into LEU targets as were contained in its previous HEU targets.[11] This enables CNEA to produce the same amount of Mo-99 with only 2 to 3 times more targets, even though LEU has only about one-fifth as much U-235 as HEU. Moreover, the new targets are compatible with CNEA's alkaline dissolution processes. The report also importantly observes that because CNEA only produces Mo-99 once a week, it had enough hot-cell down-time to develop and implement conversion without a second production line.[12] The report does not, however, go into great detail about the back-end outcomes and processes in Argentina – including Mo-99 efficiency and purity, minimization of process waste streams, and waste management.[13]

While secondary sources in the public domain offer insight into the technical challenges of conversion, the motivations for conversion are less well documented. According to one published source, CNEA made its decision based on the "final cutoff in HEU provision."[14] According to other sources, CNEA's goal was to "reduce practically to zero the total HEU inventory ... in accordance with the

RERTR objective,"[15] thereby conforming to international norms and policies.[16] But these explanations are not well developed, and leave open a range of other possible explanations.

Alan Kuperman suggests that Argentina and other countries that adopt LEU-based production could benefit from a "first mover" advantage. The logic is that if the governments of Mo-99 importing states were to enforce a preference for non-HEU-produced isotopes, the first producers to use LEU would gain long-term market share due to the "highly competitive nature of the isotope market and the premium that customers place on reliable supply once achieved."[17] The benefits that Argentina sought to gain through conversion, however, are unclear and warrant further attention.

In recent years, research has compared HEU- and LEU-based production of Mo-99. For example, Cestau and his co-authors undertook in 2007 to compare production efficiency of the two methods and concluded that LEU targets can achieve the same production efficiency as HEU targets.[18] Nonetheless, the world's largest producers so far have refused to commit the time and resources to convert, so it is worth further exploring the lessons of the Argentine case.

## Methodology

This chapter draws upon the Argentine case to identify lessons for phasing out, on a global scale, the use of HEU in medical isotope production. To do this, the chapter approaches the case study from three angles.

First, it explores why Argentina converted. This is the driving question of my research, and the aspect that is least clear from the existing literature. I aim to understand the incentives and disincentives that guided the decision to convert, and whether the decision was based on cost, security, HEU supply, or other considerations. For example, I explore whether the decision was made in response to international pressure, or in anticipation of the future trajectory of medical isotope commerce, or due primarily to domestic pressures and other practical considerations. Further, I seek to identify who served as primary sources of support and opposition, and whether the decision-making process involved public participation. Finally, I investigate what degree of regulatory burden or opposition was faced by CNEA.

Second, the chapter analyzes the technical and logistical details of Argentina's conversion to LEU targets, examining if and how CNEA was able cost-effectively to convert its Mo-99 production, to identify factors that contributed to the country's success.

Third, I explore the retrospective assessment by Argentine officials of the wisdom of their decision to convert, almost a decade after implementing it. I investigate whether their expectations of cost and time were accurate, and whether Argentina has encountered any unexpected advantages or disadvantages from conversion?

To answer these questions, I supplement my review of the existing literature with interviews of officials who were involved in the conversion process. My key interviewees in Argentina were Pablo Cristini, manager of the Radioisotope

Applications and Technology Division at CNEA's Ezeiza Atomic Center, located outside of Buenos Aires; and Hugo Vicens, a former official at Argentina's ARN. The Ezeiza Atomic Center was also kind enough to offer me a tour of its medical isotope production facilities. In the United States, I reached out to George Vandegrift at Argonne National Laboratory, and Rilla Hamilton at the Global Threat Reduction Initiative of the Department of Energy (DOE), to assess the extent of international involvement in Argentina's conversion. Underlying my entire project is a fundamental question: Why did the alleged impediments to conversion, commonly cited by the world's largest isotope producers who continue to use HEU, not deter or prevent Argentina from successfully phasing out HEU for the production of medical isotopes?

## Research Findings

### *Incentives and Disincentives for Conversion*

A review of the literature quickly makes clear that the primary concern for Mo-99 producers, and the governments that support them, is maintaining a reliable supply of medical isotopes at a competitive price. Thus, perhaps not surprisingly, interviews with Argentine officials reveal that reliability-of-supply concerns far exceeded those about nuclear security in driving the determination to convert to LEU-based isotope production.

According to the NAS report, typical concerns about reliability of supply stem from the following three factors:

1 Increase in global demand for Mo-99;
2 Reliance on aging reactors for medical-isotope production; and
3 The increased difficulty of transporting Mo-99 across international borders.[19]

The NAS report argues, however, that these three factors are not incentives to convert because even if major producers used LEU targets, the same concerns would apply. But a fourth factor cited by the above report – "long-term uncertainties associated with the continued availability of HEU for Mo-99 production"[20] – more likely drove Argentina's decision-making.

Prior to converting to LEU targets, Argentina imported HEU every two years from the United States.[21] In 1992, however, the Schumer Amendment to the Energy Policy Act mandated that for any future license to export HEU for use as fuel or target in a foreign reactor, the U.S. Nuclear Regulatory Commission (NRC) had to verify that the following conditions were met:

1 No LEU alternative yet existed that could fuel the reactor or be used as a target for isotope production.
2 The foreign operator was committed to using LEU as soon as feasible.
3 The United States was actively developing an LEU alternative suitable for the facility in question.[22]

According to Loukianova and Hansel, the 1992 Schumer Amendment played a critical role in persuading foreign operators to cooperate with the RERTR program toward converting their facilities.[23] Soon after the law's enactment, the incoming Bill Clinton administration also promoted HEU minimization through its nonproliferation policy, outlined in Presidential Decision Directive/NSC-13, in September 1993. As a result of these initiatives, the United States did not export any HEU in 1995 and 1996.[24]

Interestingly, Hugo Vicens, who at the time of conversion was employed at Argentina's ARN, is adamant that neither his organization nor CNEA was concerned about the physical protection of the HEU stored at the Ezeiza Atomic Center.[25] Rather, CNEA's objective was to complete its "social function" of ensuring reliable isotope supply.[26]

### *Support for (and Opposition to?) Conversion*

By the late 1990s, the United States had cut off exports of fresh HEU to Argentina, and officials at CNEA and in the Argentine government recognized that existing reserves were dwindling. According to Cristini, this is what drove the enormous commitment by CNEA, ARN, and concerned Argentine ministries to convert to LEU targets. In Cristini's opinion, finding consensus was never a problem. Indeed, he refers to the decision as *una cuestión patrimonial*, or a matter of patrimony. In other words, Argentina embraced conversion as a patriotic means of ensuring its ability to continue producing medical isotopes. CNEA then shored up this support for conversion by demonstrating that it was technically feasible to continue producing medical isotopes in existing facilities using LEU targets.[27]

At the outset of my research, I suspected that two other entities had vested interests and thus possible reasons to support or oppose CNEA's attempts to convert to LEU targets. The first entity is INVAP, a semi-private Argentine company that is devoted to the design and construction of complex technological systems, and that serves essentially as the trade arm of CNEA. While INVAP is an independent company, CNEA has representatives on INVAP's board of directors. In the wake of CNEA's success, INVAP has marketed CNEA's LEU-based isotope production technology to countries such as Australia and Egypt, and thus may have had reason to support the original conversion decision. But Cristini and Vicens both assured me that although INVAP engages in research and development, the company was in no way involved in the original conversion decision.

A second potential vested interest group was Argentina's nuclear medical professionals. Domestic medical practitioners could have been apprehensive about conversion due to fears of a disruption in supply or diminution in quality. Alternatively, they might have supported conversion to LEU, in light of the impending exhaustion of Argentina's stock of HEU for targets. Cristini contends that the Argentine medical community supported whatever decision CNEA would make. Thus, he says, the decision to convert ultimately was made by CNEA.[28]

### *International Incentives*

As noted, the Schumer Amendment provided the RERTR program with leverage in its international efforts to phase out civilian uses of HEU. The United States was involved in Argentina's conversion process in several key ways. First and foremost, by cutting off HEU supply, the United States forced Argentina to make tough decisions about its medical-isotope program. But the United States offered carrots as well. It made a commitment to encapsulate, transport, and repatriate the accumulation of spent, radioactive HEU waste in Argentina. Moreover, the United States offered to supply Argentina with quantities of fresh LEU that were equivalent in terms of U-235 content. Most importantly, Argonne collaborated with CNEA to develop LEU targets for irradiation in Argentina's R-3 reactor.

Still, George Vandegrift, a distinguished fellow at Argonne, stresses that CNEA took the initiative: "We worked closely with the Argentines, but their conversion efforts were based almost entirely on their own development activities. They have always been strong supporters of RERTR efforts, and as far as I know their conversion was based on internal conviction that it was the right thing to do."[29]

### *Logistics of Conversion*

Regardless of motivation, conversion also required technical and economic feasibility. Argentine officials have made clear in their presentations to RERTR conferences that conversion was not cost-prohibitive. One factor facilitating affordability is that the same set of hot cells was used for LEU-based production as had been used for HEU-based production. Indeed, the NAS study notes that the most expensive portion of a facility for processing targets is its hot cells.[30] Thus, one of CNEA's biggest challenges, and accomplishments, was utilizing its existing hot cells to develop and implement conversion to LEU targets, while continuing without interruption to produce Mo-99 with HEU targets in the same hot cells until conversion was complete.

To understand how this was possible, it is important to review how hot cells are used in isotope production, as described in the NAS report. When irradiated targets are removed from the reactor, they are cooled in water before being taken to the processing facility (in Argentina, both are on the same site) in shielded casks. The targets are then placed into the hot cells for processing to recover Mo-99. This must be done quickly to minimize loss of Mo-99 from radioactive decay. Inside the hot cell, the targets are dissolved in a container, connected to tubing and columns for subsequent chemical separations that isolate the Mo-99.[31]

According to Cristini and Vicens, Argentina has two lines of hot cells, an old line and a new one. At the time of conversion, the new line of cells was used for production and the old line was used for storage. Nuclear waste, including spent target material, was kept in the old line of hot cells to cool before being moved to dry storage.

As isotope production was being developed using LEU targets, CNEA was able to overlap HEU and LEU processes for four full cycles each during the course of a month in 2002. Cristini noted that there was enough room in the reactor to irradiate both HEU and LEU targets simultaneously, and that the targets could be distinguished because they were of two different dimensions.[32] Once the irradiated targets had been transferred to the hot cells, CNEA used separate filters for recovery of Mo-99 from HEU and LEU targets.

Since converting to LEU targets, Argentina has taken several steps to increase output of Mo-99. CNEA's need to increase output became critical in 2007 and 2008 when Canada's NRU reactor shut down, causing consumers in Latin America to look to Argentina to fill the supply gap. CNEA doubled the number of targets in each batch irradiated in the reactor at one time, and increased the reactor's power from five megawatts to between nine and 10 megawatts.[33]

According to Cristini, Argentina seeks to produce at even larger scale. But in order to do so, CNEA must reduce the downtime in production that occurs when targets are being discharged from the reactor. This is a matter of solving technological problems in the reactor.[34] CNEA is also cooperating with Argonne to develop a metal foil target that can be used in the R-3 reactor. If these efforts are successful, Cristini and Vicens say that CNEA will switch to the new targets. Metal foil targets have greater density and other properties that would increase the efficiency of production. For example, when CNEA receives LEU metal, it only would have to perform a metallurgical process to laminate the material to create a metal foil target. By contrast, to fabricate the current targets, LEU metal first must be converted to an aluminum alloy.[35]

### *Expectations Met?*

In retrospect, neither Pablo Cristini nor Hugo Vicens perceived any negative consequences from conversion. This likely is because CNEA was able to test the LEU target without disrupting production and to convert in an affordable manner. CNEA's priority was to ensure the stable production of Mo-99, and it accomplished that goal.

## Important Takeaways

Argentina has expressed a sense of pride in its conversion and its technological leadership in LEU-target design and fabrication. Indeed, Argentina's conversion enables the country to tout its commitment to HEU minimization, as part of a regional and global nonproliferation agenda. But concerns about nuclear security did not drive Argentina's conversion. This case study demonstrates that the decision to convert was fundamentally economic. Argentina faced a cutoff in supply of HEU, a key input for its Mo-99 production.

The Argentina case also demonstrates that although the United States is currently entirely dependent upon foreign producers for its supply of Mo-99, it nonetheless can exercise significant influence over the calculations by these producers of the costs and benefits of conversion. This leverage has three sources, each arising from global supply and demand chains.

First, the United States is a principle provider of HEU for targets, including to Canada, where Nordion is one of world's two largest producers of Mo-99. In the opinion of Pablo Cristini at CNEA, had the United States enforced the Schumer Amendment and discontinued HEU provision to producers who failed to prove a willingness to convert, global conversion to LEU already would have happened.[36]

Second, not only does the United States consume approximately half the global supply of Mo-99, but it represents a consistent source of demand in an industry that craves stability due to the short half-life of its product. This dominant consumer role also gives the United States leverage to persuade producers to eliminate the use of HEU. This leverage should increase as non-HEU-based suppliers of Mo-99 enter the global market, through conversion of existing producers or start-up of new entrants. The South African Nuclear Energy Corporation recently set a precedent by shipping its first commercial batch of LEU-based isotopes to the United States in July 2010.[37] Should the United States demonstrate a firm commitment to phasing out the import of HEU-based medical isotopes, large-scale producers would have strong incentive to convert to LEU-based production.

The final source of leverage stems from the U.S. willingness and ability to subsidize development of non-HEU-based isotope production technologies. The United States has already played an important role in reducing the costs of conversion by financing research and development of LEU-based production. Argonne continues to work with CNEA to develop new techniques for LEU target fabrication. Likewise, the United States recently provided a $25 million grant to assist conversion efforts in South Africa.

If phasing out the use of HEU targets in medical isotope production is a national security priority, U.S. officials should ensure that they make full use of this extensive U.S. leverage. It will require a shrewd but forceful package of policies. Some of these options are highlighted in the following section.

## Conclusions and Policy Recommendations

Any policy initiative for a global phase-out of HEU-based medical isotopes must take into full consideration the need to guarantee continuity of medical isotope supply. Avoiding supply disruptions requires market-based incentives that on the one hand encourage non-HEU-based production, but on the other hand do not increase costs of production to such a degree that they drive producers out of business. The Schumer Amendment was more or less successful in accomplishing this. That U.S. law compelled Argentina to convert to LEU targets for fear that the cutoff of supply of HEU would cause severe disruptions in the Southern Cone's medical isotope market. Wisely, by also mandating U.S. involvement in development of LEU alternatives, the amendment at the same time facilitated cooperation between U.S. government entities and CNEA to develop suitable LEU targets for Argentina.

Policy makers should keep in mind, however, that Argentina is the primary medical isotope producer in the Latin American market for reasons of proximity more than cost. In choosing to convert, Argentina did not face the same level of risk that major producers for the European and North American markets might assume by choosing to convert, because Argentina did not need to be as concerned about losing market share to lower-cost competition. Since 2007, the world's three largest producers – in Canada, the Netherlands, and Belgium – have experienced disruptions in production at some point, due to malfunctions or regulatory concerns.[38] These producers may be reluctant to experiment with LEU-based production for fear of again disrupting their production.

In light of these objectives and concerns, U.S. policy with regard to worldwide Mo-99 production should embrace the following strategies:

1 Continue to encourage collaboration between U.S. national laboratories and foreign isotope producers.
2 Provide adequate funding for research and development of alternative, non-HEU-based production methods.
3 Phase-out exports of HEU to foreign medical isotope producers by reinstituting and universally enforcing the terms of the Schumer Amendment.
4 Fund programs for U.S. production of medical isotopes without HEU.

Over the last few years, two initiatives in the U.S. Congress have attempted to enact such policy. In 2009, the U.S. House of Representatives passed bill H.R. 3276, which included provisions that reversed the Burr Amendment and restored the intent of the Schumer Amendment. The bill took steps to phase-out U.S. exports of HEU and to fund domestic production of Mo-99 without HEU. H.R. 3276 never made it past the Senate, however, because the bill's chief opponent, then-Senator Kit Bond of Missouri, successfully kept it from being brought to a vote. According to a 2010 letter from members of the U.S. nuclear medicine and nonproliferation communities to Senator Bond, the House bill included two measures to promote domestic production of Mo-99 without HEU. First, it subsidized construction of production facilities by authorizing government cost-sharing. Second, it facilitated operation of new facilities by authorizing the government "to retain responsibility for the final disposition of radioactive waste" under uranium-lease agreements – without which operators would have nowhere to dispose of this waste.[39]

In the following Congress, Senators Jeff Bingaman and Lisa Murkowski introduced an almost identical bill to revive the effort. The American Medical Isotopes Production Act of 2011 (S. 99) aimed simultaneously to improve the reliability of domestic medical isotope supply and to phase out the export of HEU for medical isotope production. The bill included language that was strikingly similar to the Schumer Amendment in outlining a program to support large-scale domestic production of Mo-99 for medical purposes without HEU. The bill barred this assistance to any domestic reactor fueled with HEU or using HEU targets unless LEU alternatives were unavailable for that reactor. Moreover, operators of such reactors needed to provide assurances that they would use LEU fuel and targets in the reactor as soon as they become available.

Concomitantly, the bill required that in seven years the NRC must discontinue export licenses for HEU for the purposes of medical isotope production. Yet, the bill recognized the need to ensure an adequate supply of medical isotopes. Accordingly, in the event that after seven years the global supply of non-HEU-based Mo-99 proved insufficient to satisfy the U.S. domestic market, and if the export of U.S.-origin HEU were the surest way to increase that supply, the bill permitted HEU exports to be extended for up to six more years.[40]

In January 2013, S.99 was signed into law. The value of its approach is three-fold. First, it gives adequate time for medical isotope producers to adapt to the new regulations. Companies need to plan for any changes in production methods, to minimize the costs of conversion and keep the price of their Mo-99 competitive and affordable. Second, in the short run, U.S. government support will help producers who avoid HEU compete with foreign companies that use HEU and are highly subsidized. Third, in the longer run, the bill helps level the playing field in an industry that is very competitive. In theory, if all producers share the same deadline for conversion, then no single producer can gain an advantage by refusing to convert. On such a level playing field, it is even possible that producers might be more willing to cooperate – for example, to coordinate their planned outages during maintenance or conversion, to avoid supply disruptions. Obviously, the effectiveness of the bill also could be enhanced if the United States worked in conjunction with the other potential HEU supplier, Russia.

The Argentine case demonstrates that the recent enactment of S. 99 – which includes a well-designed combination of forewarning, investment, and enforcement – could promote a global phase-out of HEU for production of medical isotopes. The specifics of Argentina's conversion may not be directly applicable to larger-scale producers. But if the technical capacity exists for the world's largest Mo-99 producers to convert to LEU targets, as now appears the case, then the United States can and should provide the economic incentives to foster that outcome. Based on the Argentine experience, when the economic incentive is there, the political will to convert is likely to follow.

## Notes

1 Frank N. von Hippel and Laura H. Kahn, "Feasibility of Eliminating the Use of Highly Enriched Uranium in the Production of Medical Radioisotopes," *Science and Global Security* 14 (2006): 152.

2 IAEA Staff Report, "IAEA Helps to Close Radioisotope Production Gap," 7 January 2011, http://www.iaea.org/newscenter/news/2011/prodgap.html (accessed on 27 September 2012).

3 See National Research Council, *Medical Isotope Production Without Highly Enriched Uranium* (Washington, D.C.: National Academies Press, 2009): 103; and G. F. Vandegrift, C. Conner, G. L. Hofman, R. A. Leonard, A. Mutalib, J. Sedlet, D. E. Walker, T. C. Wiencek, and J. L. Snelgrove, "Modification of Targets and Processes for Conversion of 99Mo Production from High- to Low-Enriched Uranium," *Industrial and Engineering Chemical Research* 39, 9 (2000): 3140.

4 U.S. House of Representatives, Conference Report on H.R. 776: Comprehensive National Energy Policy Act, H.R. Rep. No. 102-H12103 (1992), http://www.rertr.anl.gov/REFDOCS/EPACT92.htm (accessed on 27 September 2012).

5 The RERTR program was established in 1978 by the U.S. Department of Energy to phase out the use of HEU in research and test reactors. Vandegrift et al., "Modification of Targets," 3140, provides a concise summary of RERTR's relationship with global medical isotope commerce. For a more recent update, see Silvia Balart, Pablo Cristini, Alfredo G. Gonzalez, Rubén González, J. Daniel Hermida, Marisol López, Marcela Mirandou, and Horacio Taboada, "2009 Progress Report on RERTR Activities in Argentina," paper presented at RERTR 2009 31st International Meeting on Reduced Enrichment for Research and Test Reactors, Beijing, 1–5 November 2009, http://www.rertr.anl.gov/RERTR31/ (accessed on 27 September 2012).

6 "Country Profile: Canada," Nuclear Threat Initiative, May 2009, http://www.nti.org/db/heu/canada.html (accessed on 10 April 2011).

7 Anya Loukianova and Cristina Hansell, "Leveraging U.S. Policy for a Global Commitment to HEU Elimination," *The Nonproliferation Review* 15, 2 (July 2008): 173.

8 For more information, see CNEA's website at http://www.cnea.gov.ar/index.php (accessed on 27 September 2012).

9 For a more recent ENS update, see Silvia Balart, Herman Blaumann, Pablo Cristini, Alfredo G. Gonzalez, Rubén González, J. Daniel Hermida, Marisol López, Marcela Mirandou, and Horacio Taboada, "2010 National Progress Report on R&D on LEU Fuel and Target Technology in Argentina," paper presented at RRFM 2010, Marakhesh, Morocco, 21–25 March 2010, http://www.euronuclear.org/meetings/rrfm2010/index.htm (accessed on 27 September 2012).

10 Balart et al., "2009 Progress Report," 3–4. See also Pablo R. Cristini, Héctor J. Cols, Ricardo Bavaro, Marcelo Bronca, Roberto Centurión, and Daniel Cestau, "Production of Molybdenum-99 from Low Enriched Uranium Targets," paper presented at RERTR 2002, San Carlos de Bariloche, Argentina, 3–8 November 2002, http://www.rertr.anl.gov/Web2002/2003Web/Cristini.html (accessed on 27 September 2012).

11 National Research Council, *Medical Isotope Production,* 95.

12 National Research Council, *Medical Isotope Production,* 96, 109.

13 The National Research Council identifies these back-end processes as key issues in the conversion process on p. 98 of its report.

14 Balart et al., "2009 Progress Report."

15 L. Alvarez, N. Boero, P. Echenique, E. Pasqualini, G. Ruggirello, and H. Taboada, "Status of RERTR Activities in Argentina," paper presented at RERTR 2003 25th Annual International Meeting on Reduced Enrichment for Research and Test Reactors, Chicago, 5–10 October 2003, Argonne National Laboratory, http://www.rertr.anl.gov/RERTR25/index.html (accessed on 27 September 2012).

16 Daniel Cestau, Ariel Novello, Pablo Cristini, Marcelo Bronca, Roberto Centurión, Ricardo Bavaro, Julián Cestau, and Eduardo Carranza, "HEU and LEU Comparison in the Production of Molybdenum-99," paper presented at RERTR 2007, Prague, Czech Republic, 23–27 September 2007, http://www.rertr.anl.gov/RERTR29/Abstracts/S6-4_Cestau.html (accessed on 27 September 2012).

17 A. J. Kuperman, "The Global Threat Reduction Initiative and Conversion of Isotope Production to LEU Targets," paper presented at RERTR 2004, Vienna, Austria, 7–12 November 2004, http://www.rertr.anl.gov/RERTR26/Abstracts/17-Kuperman.html (accessed on 27 September 2012).

18 Cestau et al., "HEU and LEU Comparison."

19 National Research Council, *Medical Isotope Production,* 55.

20 National Research Council, *Medical Isotope Production*, 63.

21 Hugo Vicens, interview by author, Ezeiza Atomic Center, Ezeiza, Buenos Aires, Argentina, 11 January 2011.

22 U.S. House of Representatives, Conference Report on H.R. 776.

23 Loukianova and Hansel, "Leveraging U.S. Policy," 173.
24 Loukianova and Hansel, "Leveraging U.S. Policy," 165.
25 Interview with Hugo Vicens, 11 January 2011.
26 Pablo Cristini, interview by author, Ezeiza Atomic Center, Ezeiza, Buenos Aires, Argentina, 11 January 2011.
27 Interview with Cristini.
28 Interview with Cristini.
29 George F. Vandegrift, email interview by author, 4 January 2011.
30 National Research Council, *Medical Isotope Production*, 25.
31 National Research Council, *Medical Isotope Production*, 25.
32 Interview with Pablo Cristini, 11 January 2011.
33 Interview with Hugo Vicens, 11 January 2011.
34 Interview with Pablo Cristini, 11 January 2011.
35 Interview with Hugo Vicens, 11 January 2011.
36 Interview with Pablo Cristini, 11 January 2011.
37 Miles A. Pomper and William C. Potter, "Medical Isotope Production: The US Must Follow South Africa's Lead," *Bulletin of the Atomic Scientists*, 17 December 2010, http://www.thebulletin.org/web-edition/features/medical-isotope-production-the-us-must-follow-south-africas-lead (accessed on 27 September 2012). For further detail, see this book's chapter by Chloe Colby.
38 National Research Council, *Medical Isotope Production,* 58.
39 Letter to Senator Kit Bond from Alan J. Kuperman et al., 20 July 2010, http://www.utexas.edu/lbj/sites/default/files/file/news/Isotopes-Bond-group-letter-2010-July-Final-3.pdf (accessed on 27 September 2012).
40 American Medical Isotopes Production Act of 2011, S.99 , 112th Cong. (2011), http://www.govtrack.us/congress/bills/112/s99 (accessed on 27 September 2012).

# 3 South Africa

## Reactor Fuel and Medical Isotope Production

*Chloe Colby*

This chapter investigates the ongoing effort of the South African Nuclear Energy Corporation (NECSA) to phase out use of highly enriched uranium (HEU), in favor of low-enriched uranium (LEU), in the targets it uses to produce molybdenum-99 (Mo-99), the key ingredient in the world's most commonly used medical isotopes. The targets are small and easy to transport and therefore – when composed of HEU, a nuclear weapons material – an obvious proliferation and terrorism risk. In contrast to the HEU targets, NECSA had by 2009 already completed converting the fuel of its reactor, the South African Fundamental Atomic Research Installation (SAFARI-1), from HEU to LEU. More recently, NECSA's subsidiary company, NTP Radioisotopes, has pioneered conversion to LEU targets for the large-scale, commercial production of Mo-99. Previously, LEU targets had been used only for small-scale Mo-99 production – in Australia and Argentina.[1]

The rest of this chapter explores the technical, economic, and political obstacles that NTP encountered, and how they were overcome. It also investigates whether, somewhat counter-intuitively, there may be some economic incentive for conversion to LEU targets. It finds that after the primary technical obstacles were surmounted, the most difficult ongoing struggles for NTP have been (1) the increased cost of importing LEU targets from France, and (2) the recently depressed world price of Mo-99 due to decreased demand following a 2009 supply shortage that shocked the global market. The original political impetus for NTP's switch to LEU came from international actors, who successfully encouraged the South African government to mandate conversion. Since then, the domestic producer has played the main role in overcoming obstacles to implementation.

Financial and technical support for conversion has come from the United States. This includes the National Nuclear Security Administration (NNSA) in the Department of Energy, whose funding has been instrumental in the development and implementation of the target conversion, and the Reduced Enrichment for Research and Test Reactors (RERTR) Program at Argonne National Laboratory. After reviewing the challenges, incentives, and accomplishments in the South African case, this chapter concludes with recommendations to encourage a global phase-out of HEU for the production of medical isotopes.

## Literature Review

SAFARI-1 is a 20 MW tank-in-pool type, light water reactor – based on the U.S. Oak Ridge Reactor design – that has been operating at the NECSA site at Pelindaba since going critical on 18 March 1965. The fuel's source and enrichment level have varied over the reactor's lifetime due to changes in the political environment.[2] The reactor was initially fueled by HEU from the United States and then, from 1981 until recently, used HEU from South Africa's own inventory, enriched to either 45% or 90% U-235.[3] South Africa's Department of Minerals and Energy approved the conversion of the SAFARI-1 core to LEU fuel in July 2005.[4] As of 25 June 2009, the reactor's core has been fully fueled with LEU.[5]

The RERTR program holds annual conferences that have produced a significant literature on converting research reactors to use LEU fuel and, more recently, converting Mo-99 production to LEU targets.[6] South Africa's conversion to LEU targets is quite recent, so the literature focuses instead on the conversion of its reactor fuel to LEU, and the government's growing support for international nonproliferation policy since the country acceded to the Nuclear Non-Proliferation Treaty in 1991.[7] In addition, the broader literature on worldwide efforts to eliminate HEU from Mo-99 production has at times speculated on NTP's potential to convert.

The key commercial activity at SAFARI-1 is NTP's production of Mo-99.[8] Moreover, NTP is one of only four large-scale producers in the world, and the company supplies about 10 percent of the global demand of Mo-99.[9] The isotope is produced by irradiating in the reactor small targets containing U-235, which fissions to produce Mo-99, which is then extracted by chemically processing the irradiated targets. NTP traditionally has used targets fabricated from a supply of South African-produced HEU enriched to 45% U-235.[10] By contrast, the HEU targets used in most of the world's Mo-99 production are 93%-enriched.

Argentina's prior experience illuminates some of the technical and economic challenges of conversion. (For additional details, see this book's chapter by Jared Berenter.) Since 2002, Argentina's Comisión Nacional de Energía Atómica (CNEA) has successfully used LEU targets to produce Mo-99 in its RA-3 reactor, reporting it to be as efficient and reliable as before conversion.[11] CNEA previously used 93%-enriched, HEU targets, composed of a uranium-aluminum alloy with a uranium density of about 1.6 g/cm$^3$.[12] To make an LEU target of 19.9 percent enrichment, with the same volume of target meat and mass of U-235, would require a uranium density of approximately 7.5 g/cm$^3$. But such a high-density is not technically feasible with Argentina's LEU target, composed of a uranium-aluminum dispersion. Accordingly, CNEA took three steps to squeeze more U-235 into its LEU target, which is about the same size as the preceding HEU version: the density of the target was increased from 1.6 g/cm$^3$ to 2.9 g/cm$^3$; the target meat thickness was increased; and the thickness of the cladding was decreased. However, because the uranium density was not even doubled, the LEU target still has considerably less U-235 than the original HEU target. As a result, CNEA must irradiate two to three times more LEU targets to produce the same quantity of Mo-99.

In South Africa, by contrast, NTP's HEU targets were only 45%-enriched. So, the company already had been processing twice as many targets as other producers who use 93%-enriched uranium targets to produce the same amount of Mo-99.[13] This made it easier technically for NTP to convert to LEU targets without hindering output.

Economic considerations, however, would play a significant role in determining the viability of NTP's conversion, because as noted Mo-99 production is the main commercial activity of the SAFARI-1 reactor. In addition, any potential disruption of production during conversion would have global ramifications, because the world has so few large-scale isotope suppliers. This was amply demonstrated in 2009, when the simultaneous shutdown of two major isotope-production reactors – the NRU in Canada, and the HFR in the Netherlands – created a worldwide shortage of Mo-99.[14]

A stable supply of medical isotopes is a major concern for the United States and most other industrialized countries. Indeed, the United States alone accounts for about half of the world's demand for medical isotopes. Accordingly, the U.S. Department of Energy for several years has provided cost sharing to help establish domestic production of Mo-99 without HEU. In January 2013, the United States enacted associated legislation to authorize such cost sharing and phase out HEU exports, the American Medical Isotopes Production Act.[15]

Building on the above published information, this chapter attempts to determine the status of NTP's conversion, detailing the company's specific experience. The goal is to understand more fully the challenges and incentives of commercial-scale isotope production with LEU targets, in order to develop policies to encourage other commercial isotope producers to phase out the use of HEU.

## Methodology

My research examined three main questions: (1) the reasons behind NECSA's decision to convert to LEU targets; (2) the technical, economic, and political obstacles it encountered in the process; and (3) the current status of the conversion. I explored whether there was initial resistance to conversion, how this resistance and other obstacles were overcome, and what NECSA is currently doing to facilitate conversion. Due to the global nature of the efforts to phase out HEU, I also investigated international pressures and incentives.

The most effective method for answering these questions was personal interviews with NTP officials. The topic is so recent and ongoing that even the few published reports have quickly become outdated. Following conversion of the reactor's fuel, journalistic coverage dwindled and relatively little attention has been paid since to the conversion of targets. In addition to interviews in South Africa, my primary sources included informal discussions with officials from the U.S. Department of Energy's Global Threat Reduction Initiative (GTRI), email correspondence with NTP and NECSA officials, and unpublished documents including an internal NECSA report and some intra-industry materials. Secondary sources included recent news articles and press releases.

The individuals most helpful with this research were the NECSA and NTP employees who experienced the fuel conversion and have worked on the target conversion. It is NECSA, not the South African Department of Energy, which takes the lead on conversion and is GTRI's main point of contact. Dr. Charles Piani was NECSA's Senior Manager in charge of the SAFARI-1 reactor throughout the conversion of the core. He wrote the most relevant papers, which he presented at several RERTR conferences, and has been involved with the South African nuclear community since the 1970s. Although he is not involved in the current target conversion efforts, he was instrumental in directing me to those who are. NTP's Chief Technical Officer, Dr. Gavin Ball, has been part of the process from the beginning, but was constrained in answering technical questions due to the proprietary nature of the information. Mr. Roland von Gogh is the Marketing and Sales Manager for NTP and was of invaluable assistance in facilitating contacts within NTP, as well as discussing economic and political issues associated with the target conversion. Several other NTP and NECSA officials provided additional correspondence. Ms. Gugu Dube of the Institute for Security Studies in Pretoria was also consulted.

## Findings

### *Technical Challenges*

The biggest hurdle for conversion of HEU targets was an initial technical problem that had economic implications: namely, inconsistency in the amount of Mo-99 that the LEU targets were producing, a problem known as "variable yield." According to an NTP official, there was no inherent resistance to the idea of converting, but there was concern about the effects on Mo-99 yields. NECSA CEO Rob Adam revealed in 2010 that the target conversion had encountered "significant technical hurdles" including variable yield.[16] Such variation was a major worry because NTP had structured its business around the assumption that targets would yield 80 percent of the theoretical maximum amount of extractable Mo-99. But this concern has now been successfully addressed, according to NTP's presentation at the RERTR meeting in October 2011:

> After the completion of the development work to optimize the Mo-99 extraction efficiencies, the yields obtained from the LEU Mo-99 production runs are equivalent to the HEU Mo-99 production runs. It must, however, be pointed out that the batch sizes have decreased by approximately 10 percent due to the U-235 loading in the LEU targets being lower. This is as a result of uranium density restrictions [currently 2.75 g/cc] in the dispersed target.[17]

In 2009, NTP began using LEU targets to produce some Mo-99, also known as "LEU-LEU Mo-99" because both the reactor fuel and target were low-enriched. In July 2010, the United States received its first batch of Mo-99 from South African produced exclusively with LEU targets.[18] However, NTP will keep producing some Mo-99 with HEU targets until LEU targets have proven to be a stable and reliable source of isotopes, and of income for the company. By late 2010, LEU targets still accounted for only 10 percent of NTP's Mo-99 production. That share grew to 35 percent by the second half of 2011.[19]

Total conversion to LEU targets was scheduled for late 2011, but has been postponed. The main reason appears to be that health authorities in Europe, where NTP sells some Mo-99, have not yet licensed the company's LEU-produced version, so premature conversion by the company would sacrifice some of its sales and market share. The company also cites several other obstacles to quickly increasing the percentage of LEU-based production: it is less efficient than HEU-based production and will be until more U-235 can be squeezed into the LEU target; it can interfere with HEU-based production; the LEU targets are not yet qualified to produce another commercial isotope, I-131; and customers do not yet demand LEU-based Mo-99.[20]

NTP's LEU dispersion targets are supplied by CERCA, a branch of France's AREVA. But NTP may have influenced their design. According to correspondence with an NTP official, the company has "introduced subtleties and modifications to suit the purpose of Mo-99 production on a daily basis at commercial scale."[21] These subtleties and modifications were not elaborated.

The increase in radioactive waste inherent in producing with LEU targets is also a "serious technical challenge." No further details were provided due to proprietary concerns. NTP does not patent technology, including for waste disposal or modifying CERCA's LEU targets, so it guards such information as a trade secret. NTP is in the process of licensing the target manufacturing technology from CERCA, and NECSA is building new facilities to begin producing targets domestically using LEU blended down from South Africa's own HEU stockpile.

NTP's embrace of conversion has crucial implications for global isotope production. The initial, successful runs of LEU targets have proved that a large Mo-99 production facility can viably and safely continue its output of Mo-99 after conversion to LEU reactor fuel and LEU targets. NTP has not decided whether this technology will be made available to others on a commercial basis. But if other isotope producers are interested, NTP has not ruled out the possibility of contracting out its technology, and the company does hope eventually to export its LEU targets for irradiation elsewhere.

### *Economic Challenges*

As opposed to some political incentives to convert, NTP sees only economic disincentives. The LEU targets must be imported at a cost much higher than that of the HEU targets that NECSA manufactures domestically. According to NTP's Roland von Gogh, "price pressures have increased and [NTP's] margins have come under attack because LEU Mo-99 is at least 10 percent more expensive to produce than HEU Mo-99. There is also currently a global oversupply of Mo-99 which is putting downward pressure on prices."[22] Although most press coverage, and even NECSA's own 2010 Annual Report, focuses on the global shortage of

Mo-99 in 2009, there is now actually a global oversupply of Mo-99, according to NTP's most recent sales figures. The primary reason is that during the 15 months that Canada's NRU was shut down and the world experienced an extreme Mo-99 shortage, the medical field adopted alternative procedures that use Mo-99 more efficiently, reducing demand by 20 percent.[23] When NRU came back on line, world supply quickly returned to its original level, but global demand was still depressed. Since then, demand has started to increase, but only gradually.

The reduced price of Mo-99 threatens NTP's business model. According to NTP, the price was expected to be around $1000 per six-day curie but has dropped to $700. NTP is on "very shaky ground" if the price goes under $500 per six-day curie.[24] With the cost of producing Mo-99 from LEU targets at least 10 percent higher than from HEU targets, NTP can only hope that its customers are willing to pay a premium for nonproliferation-friendly isotopes – a preference that so far is rare outside the United States. Illustrating this, an NTP official reported that the company was in a "price skirmish" with its competitor Nordion, accusing that Canadian isotope producer of undercutting NTP's prices in order to attract customers who are driven by cost rather than concern for global nuclear security. In early 2011, NTP said it perceived the United States as the only country willing to pay more for LEU-produced isotopes. Confirming that impression, in June 2012, the U.S. government announced a plan to pay higher reimbursements for medical isotopes produced without HEU.[25] As the world's largest consumer of Mo-99, the United States can help ensure NTP's profitability despite the company's marginally higher costs for using LEU targets in the production process.

### *Political Challenges*

According to South African officials involved in both the fuel and target conversions, the government is motivated heavily by its adherence to the Nuclear Non-Proliferation Treaty and associated norms, including the minimization of civilian use of HEU. The Department of Minerals and Energy, which originally mandated the conversion of the fuel and approved the plan, no longer exists, and the South African Department of Energy is not particularly involved in the conversion process. Despite that, NECSA and NTP officials see it as their political duty to carry out the conversion of Mo-99 production.

In other countries, commercial-scale isotope producers have cited technical and economic grounds for rejecting or postponing conversion to LEU targets. But NTP has shown that the technology exists and works and that the company's profit margins have not been too adversely affected. Moreover, an NTP official who requested anonymity argued that, in contrast to the small economic disincentives, there "are only political advantages and incentives." By phasing out the use of HEU, South Africa and NECSA are not only adhering to international norms, but establishing themselves as global leaders on nonproliferation policy. The same official attributed much of the company's success to the effectiveness of NECSA CEO Rob Adam as a "scientist-politician," who has the ability to lead in both the scientific and political realms and who is truly dedicated to the conversion process.

The United States has supported and promoted NTP's conversion, including in October 2010, when the NNSA awarded NECSA a $25 million grant.[26] Both the NNSA and NTP agree that the grant was awarded to help perfect technology for producing Mo-99 with LEU targets, but their perspectives diverge somewhat beyond that point. NNSA's official position is that the grant was awarded as part of its "commitment to developing a sustainable means of producing Mo-99 as part of a global supply network that avoids a single point of failure and does not use HEU."[27] In NTP's view, the grant was awarded not only to facilitate improvements in the efficiency and capacity of LEU-based production, but to promote "the development of NTP's role in the U.S. isotope market and participation in the U.S. program to establish [domestic] Mo-99 production facilities."[28] NTP did pledge to help ensure the United States a reliable supply of Mo-99 by earmarking a significant amount of NTP's eventual output of LEU-based Mo-99 (potentially up to 4,000 six-day curies per week) for the U.S. market.[29] The NNSA also facilitated the expedited procedure by which the U.S. Food and Drug Administration (FDA), after just a few months of review, approved NTP's LEU-based Mo-99 for use in U.S. patients. This not only smoothed the way for NTP but, by eliminating an alleged roadblock, aimed to encourage other global producers of Mo-99 to convert likewise.

NTP also viewed the proposed American Medical Isotopes Production Act as another political incentive to convert. Well before the United States enacted the bill in January 2013, South African officials say it signaled a U.S. intention to move away from dependence on imported isotopes made with HEU. NECSA CEO Rob Adam raised eyebrows internationally in late 2010, when he was quoted as saying that he expected the United States to erect "some kind of tariff barrier which would make it not as attractive to import molybdenum[-99] ... derived from highly-enriched uranium,"[30] even though such an approach is not included in the U.S. legislation. An official at another leading medical isotope producer said the comment was "tantamount to a declaration of war."[31] Adam subsequently modified his statement, saying that he was only speculating and had no evidence to support such a claim.[32]

## Summary of Findings

The technical and economic issues facing commercial isotope producers looking to convert to LEU targets are substantial, but not insurmountable. Admittedly, some of the technical advances made by NTP, based on CERCA's LEU dispersion targets, are proprietary. But the commercial shipment in July 2010 of a batch of all LEU-LEU isotopes, from South Africa to the United States, proves that the technology works. Moreover, NTP aims to spread the security benefits by exporting its LEU targets to other producers in the future.

Although the existing literature warns of Mo-99 shortages, NTP reports a more recent global Mo-99 oversupply and resulting decrease in prices. LEU targets increase production costs by about 10 percent, but several actions are being taken to reduce that economic penalty. After NTP licenses CERCA's technology and starts manufacturing LEU targets in South Africa by blending down its existing HEU stockpile, costs should decrease, reducing the financial penalty of conversion. Moreover, a gradual shift in customer preferences from "cheap Mo-99" (produced with HEU) to "safe Mo-99" (produced with LEU) may further alleviate economic pressures on NTP.

The United States has taken several steps to encourage and facilitate conversion. Providing funding, as the NNSA did for NECSA with its $25 million grant, encourages technological innovation and increases incentives for large-scale producers to convert. Purchasing LEU-produced isotopes as the United States has done from South Africa, and providing higher reimbursement for non-HEU-based isotopes as the U.S. government has proposed, provide further incentives for producers to phase out the use of HEU.

## Lessons and Recommendations

NECSA's experience with phasing out HEU is inspirational for nonproliferation advocates and serves as a beacon for other isotope producers around the world who still use HEU. The success in South Africa shows that using HEU in the commercial production of medical isotopes is not necessary, the technical obstacles to LEU production can be overcome with innovation, the international political landscape is beginning to encourage conversion, and the marginal economic burden is not significant enough to deter isotope producers who are dedicated to global nuclear security.

NTP's conversion also serves as an indicator to the U.S. government that its nonproliferation policies are on the right track and that international pressure and monetary assistance are effective incentives for convincing isotope producers to convert. But this is not reason to become complacent and hope existing policies continue to work. Rather, it is a signal to increase policy efforts and stay vigilant in order to facilitate further conversion globally. The enactment of legislation such as the American Medical Isotope Production Act sends a message to the international nuclear community that phasing out the use of HEU is a priority for the U.S. government and that other countries will be expected to conform to international nonproliferation norms.

NTP's relatively expeditious conversion also highlights the foot-dragging and lack of initiative among the world's other large-scale isotope producers. If Mo-99 production is to become completely free of HEU, it is the responsibility of the producers to make the switch. NECSA is an example of a company seemingly acting against its own short-term economic interest to serve the interest of the international community, and an example of the way that other isotope producers should view their obligation to a safe, sustainable future.

It is incumbent upon the international community – not only countries that host the production facilities, but also those that import the resulting Mo-99 – to declare a preference for safer nuclear products and invest in nuclear security by supporting companies like NTP that provide alternatives to the use of HEU. Customers ultimately have the ability to determine the products that are on the market and their ingredients. Isotope-consuming countries should follow the U.S. example by implementing policies that prohibit or penalize the trade of isotopes produced using HEU and that provide incentives for the production of isotopes without HEU.

Further research should investigate the technical particulars of NTP's target modifications and address the issue of waste disposal. It also would be helpful to survey the policies that isotope-consuming countries are putting in place to incentivize conversion.

In conclusion, this chapter recommends the following policies:

- The U.S. government should continue its trend toward responsible purchase of radioisotopes by enacting legislation that discourages the use of HEU in the production process through economic measures such as subsidies for isotopes produced without HEU and tariffs on those produced with HEU.
- Foreign governments, especially in the European Union, should pursue similar legislation that financially supports conversion to non-HEU production processes for their large-scale isotope producers.
- The U.S. Department of Energy should increase funding for technological innovation in the field of medical isotope production through grants for research and development at institutions, including universities and isotope producers, which are committed to phasing out HEU.
- Medical isotope distributors, such as Lantheus Medical Imaging and Covidien, should establish a preference for procuring Mo-99 produced without HEU by committing to purchase it first when available.
- Large-scale isotope producers should make eliminating the use of HEU their top priority, consistent with an assured supply of Mo-99.

South Africa's experience demonstrates that large-scale medical isotope production does not require the use of HEU targets and that assistance from the United States can play an important role in facilitating the conversion to LEU targets. Through dedication to nuclear security, South Africa and the United States have decreased the chance of nuclear terrorism and paved the way towards a global elimination of HEU for the production of medical isotopes.

## Notes

1 George Vandegrift, "HEU vs. LEU Targets for 99Mo Production – Facts and Myths," presented on behalf of Argonne National Laboratory to the Technical Workshop, Symposium on the Minimization of Highly Enriched Uranium in the Civilian Sector, 18 June 2006, Oslo, Norway.

2 R. W. Jamie and A. Kocher, "Fuel Manufacture in South Africa: The Road to Nuclear Conversion." 2006 RERTR Meeting, Cape Town, South Africa, 29 October–2 November, 2006, p. 2.

3 Charles S. B. Piani, "SAFARI-1: Adjusting Priorities During the LEU Conversion Program." 2006 RERTR Meeting. Cape Town, South Africa, 29 October–2 November 2006, p. 1.

4 Daniel Horner and Ann MacLachlan. "SAFARI-1 Reactor to Use LEU in Landmark Conversion." *Nucleonics Week* 46, 29 (1995): 16.

5 NECSA, "Nuclear Reactor Uses Only Low Enriched Uranium (LEU) for the First Time," Media Release, 29 June 2009.
6 Gavin Ball, "Overseas Visit Report: RERTR 2010," Internal NTP Report No. NTP-REP-10/185, 3.
7 Abdul Samad Minty "South African Perspectives on Highly Enriched Uranium (HEU)," Statement at the International Symposium on Highly Enriched Uranium, Oslo, Norway, 19–20 June 2006, 1.
8 W.E. Stumpf, A.P. Vermaak, and G. Ball, "Key Considerations in the Conversion to LEU of a Mo-99 Commercially Producing Reactor: SAFARI-1 of South Africa," 2000 RERTR Meeting, Las Vegas, Nevada, USA, 1–6 October 2000, 3.
9 National Research Council, *Medical Isotope Production Without Highly Enriched Uranium* (Washington, D.C.: The National Academies Press, 2009), p. 34.
10 Piani, "SAFARI-1: Adjusting Priorities," 1.
11 Daniel Cestau, Ariel Novello, Pablo Cristini, Marcelo Bronca, Roberto Centurión, Ricardo Bavaro, Julián Cestau, and Eduardo Carranza, "HEU and LEU Comparison in the Production of Molybdenum-99," 2007 RERTR Meeting, Prague, Czech Republic, 23–27 September 2007, 6–7.
12 National Research Council, *Medical Isotope Production*, 96.
13 National Research Council, *Medical Isotope Production*, 96–98.
14 Rebecca Trager, "Medical Isotope Shortage Concerns in US," Royal Society of Chemistry, 25 November 2009, http://www.rsc.org/chemistryworld/News/2009/November/25110902.asp, (accessed on 27 September 2012).
15 "American Medical Isotopes Production Act," govtrack.us, http://www.govtrack.us/congress/bills/111/hr3276 (accessed on 27 September 2012), and http://www.govtrack.us/congress/bills/112/s99 (accessed on 27 September 2012).
16 "South African Radioisotope Production on Target," *World Nuclear News*, 17 September 2010, http://www.world-nuclear-news.org/RS-South_African_radioisotope_production_on_target-1709107.html (accessed on 27 September 2012).
17 G. Ball, O. Knoesen, and A. Kocher, "Status Update on Conversion to LEU Based 99Mo Production in South Africa," RERTR annual meeting, Santiago, Chile, 23–27 October 2011, http://www.rertr.anl.gov/RERTR33/pdfs/S4-P5_Ball.pdf (accessed on 27 September 2012).
18 "U.S. Receives LEU-Produced Medical Isotopes," *Global Security Newswire*, http://www.nti.org/gsn/article/us-receives-leu-produced-medical-isotopes/ (accessed on 27 September 2012).
19 G. Ball, "Status Update on Conversion to LEU Based 99Mo Production in South Africa," RRFM annual meeting, Prague, Czech Republic, 18–22 March 2012, 4.
20 Ball, "Status Update," 4–5.
21 Official requests anonymity.
22 Roland von Gogh, personal interview, Pelindaba, South Africa, 14 January 2011.
23 Thomas H. Maugh II, "Isotope Shortage Means a Healthcare Crisis – Los Angeles Times," *The Los Angeles Times*, 9 August 2009, <http://articles.latimes.com/2009/aug/09/science/sci-isotope9> (accessed on 27 September 2012).
24 von Gogh, personal interview, Pelindaba, South Africa, 14 January 2011.
25 "Fact Sheet: Encouraging Reliable Supplies of Molybdenum-99 Produced without Highly Enriched Uranium," The White House, Office of the Press Secretary, 7 June 2012, http://www.whitehouse.gov/the-press-office/2012/06/07/fact-sheet-encouraging-reliable-supplies-molybdenum-99-produced-without- (accessed on 27 September 2012).
26 "Necsa Consortium Celebrates $25m Contract," *World Nuclear News*, 29 October 2010, http://www.world-nuclear-news.org/RS-Necsa_consortium_celebrates_25m_award-2910107.html (accessed on 27 September 2012).
27 "Necsa Consortium Celebrates $25m Contract."
28 von Gogh, personal interview, Pelindaba, South Africa, 14 January 2011.

29 The Australian Nuclear Science and Technology Organization will also participate in some of the work that the NNSA award will fund and may eventually contribute to the Mo-99 earmarked for the United States.

30 Benjamin Humphrey, "S.African innovation fuels nuclear medicine safely," *Agence France Presse*, 12 November 2010, http://www.google.com/hostednews/afp/article/ALeqM5iBMA6Yff_tym1HQE7fyQkDF3Q_sw?docId=CNG.4ec41186499555541d520f7116297c27.161 (accessed on 27 September 2012).

31 Humphrey, "S.African innovation fuels nuclear medicine safely."

32 Interview with an NTP official conducted by the author, 14 January 2011, Pelindaba, South Africa.

# Part III

# Hopeful Progress

# 4 USA and Europe

## High-Power Research Reactors

*Margarita Jimenez*

The number of research reactors around the world fueled with highly enriched uranium (HEU) – a nuclear weapons-usable material – has declined significantly due to successful global efforts by the U.S.-based, Reduced Enrichment for Research and Test Reactors (RERTR) Program. In the United States and Europe, most of the nuclear reactors that once used HEU fuel have been converted to low enriched uranium (LEU) fuel or shut down. Only nine HEU-fueled, civilian steady-state research reactors in the United States and Europe still remain in operation. This chapter reports the progress in converting those remaining HEU-fueled nuclear reactors and details the latest research and development of ultra-high-density LEU fuels to enable conversion. Seven of the nine reactors in this study are scheduled for conversion to LEU fuels. But it will take a unified, international effort to compel the remaining two reactors to convert.

Traditionally, HEU-fueled, high-performance research reactors (HPRRs) have had smaller cores than was possible for LEU-fueled research reactors of equal power. This creates a fundamental technical challenge: how to convert HPRRs from HEU to LEU fuel without substantially enlarging the size of the core or reducing the reactor's performance? To do so, engineers must develop ultra-high-density LEU fuel, which can maintain a comparable amount of Uranium-235 in the same-sized core, permitting the reactor to continue conducting essentially the same experiments and activities.

Previous LEU fuels with uranium density up to 4.8 g/cc that have been qualified by the RERTR program – although higher in density than traditional HEU fuel of approximately 1.5 g/cc – are still too low in density to enable conversion of HPRRs without modifying the core size. But enlarging the core is impossible for some reactors and prohibitively expensive for others. Therefore, successful development of ultra-high-density LEU fuel is necessary for, and should enable, converting all or most HPRRs.

Two main fuel technologies are being pursued for the ultra-high-density LEU fuel. The first is "dispersion" fuel in which the fuel meat is a uranium-molybdenum (UMo) alloy dispersed in a solid chemical matrix. This enables uranium density of roughly 8 g/cc. The second approach is monolithic fuel in which the fuel meat is entirely, or almost entirely, a UMo alloy, enabling density above 15 g/cc.

The RERTR program aims to ensure multiple performance goals when reactors are converted from HEU to LEU fuel: the range of experiments is not significantly reduced; the lifetime of the fuel assemblies is comparable; and the cost of the new fuel is not much higher. The LEU fuel assemblies must also satisfy all safety criteria for use in the reactor core and be commercially available.[1] Conversion to LEU thus requires both fuel development and fuel fabrication capability.

The next section of this chapter reviews the previous literature on the topic. Following that, the chapter presents the research questions and methods used to answer them. Evidence includes primary sources, such as interviews, and secondary sources available to the public. The chapter next presents the findings – including the status of LEU fuel development, fabrication, testing, and conversion. The chapter concludes with policy recommendations.

## Literature Review

Five high-performance, civilian, steady-state research reactors in the United States still use HEU fuel, as indicated in Table 4.1. (This excludes non-steady-state research reactors, such as critical facilities and pulsed reactors, which have lifetime cores and thus never require fresh fuel.) The two that use the largest amount of HEU per year are the Advanced Test (ATR) Reactor at Idaho National Laboratory and the High Flux Isotope (HFIR) Reactor at Oak Ridge National Laboratory.[2] The three other U.S. HPRRs are the Missouri University Research Reactor (MURR), the National Bureau of Standards Reactor (NBSR), and the MIT Reactor-II (MITR).

In Europe, four high-performance, steady-state research reactors operate using HEU fuel, as indicated in Table 4.2. Belgium's BR-2 and France's HFR, at the *Institut Laue-Langevin* (ILL), have agreed to convert and are awaiting the development of higher-density LEU fuels.[3] Two other reactors, France's Orphée and Germany's FRM-II, also still use HEU fuel.[4]

*Table 4.1* Five U.S. High-Performance Research Reactors Still Require HEU Fuel

| *Reactor* | *Location* | *HEU kg/year* |
|---|---|---|
| Advanced Test Reactor (ATR) | Idaho National Laboratory | 120 |
| High Flux Isotope Reactor (HFIR) | Oak Ridge National Laboratory | 85 |
| Missouri University Research Reactor (MURR) | University of Missouri in Columbia | 24 |
| National Bureau of Standards Reactor (NBSR) | National Institute of Standards and Technology | 13 |
| MIT Reactor-II (MITR) | Massachusetts Institute of Technology | 8 |

Source: Daniel M. Wachs, Curtis R. Clark, and Randall J. Dunavant, "Conceptual Process Description for the Manufacture of Low-Enriched Uranium-Molybdenum Fuel," U.S. Department of Energy, Office of Nuclear Nonproliferation, Idaho Falls, February 2008, p. 4.

*Table 4.2* Four European High-Performance Research Reactors Still Require HEU Fuel

| *Reactor* | *Country* | *HEU kg/year* |
|---|---|---|
| HFR-ILL | France | 40 |
| FRM-II | Germany | 32 |
| BR-2 | Belgium | 29 |
| Orphée | France | 16 |

Sources: S. C. Mo and J. E. Matos, "Conversion Feasibility Studies for the Grenoble High Flux Reactor," Proceedings of the XIIth International RERTR Meeting, Berlin, Germany, 10–14 September 1989. Alexander Glaser and Frank von Hippel, "On the Importance of Ending the Use of HEU in the Nuclear Fuel Cycle: An Updated Assessment," 2002 International RERTR Meeting, San Carlos de Bariloche, Argentina, 3–8 November 2002. See also this book's chapter by Kenneth Dayman.

Development of ultra-high-density LEU fuel in the United States is led by Idaho National Laboratory, which pursues both dispersion and monolithic fuel. In Europe, the LEONIDAS program develops dispersion fuel through a collaborative effort between four entities: the French public company CEA; the French multinational firm AREVA that owns the French fuel fabricator CERCA; the joint European research center ILL in France; and the Belgian research center SCK-CEN.[5] LEONIDAS aims to enable the conversion of European research reactors.[6]

More is published about U.S. than European HPRRs. Existing publications also focus more on the technical aspects of fuel development, and the feasibility of conversion, than on the economic or political challenges of implementing conversion. Typically, the literature does not document the collaborative national and international efforts between various national laboratories, private companies, and reactor operators to develop new fuels and fabrication technologies. Yet, these partnerships are essential for sharing technical expertise in fuel development, as well as for promoting the shared normative objective of phasing out non-weapons uses of HEU.

## Research Questions

Each nuclear reactor using HEU fuel faces unique challenges to convert to LEU fuel. Although the fuel meat may be similar or identical for multiple reactors, each reactor requires fabrication of customized LEU fuel assemblies to maintain optimum performance. In addition, the ideal conversion process must adhere to safety regulations, minimize downtime needed for conversion, and avoid substantial cost increases.

To explore progress on converting HPRRs to LEU fuel, I posed the following questions to operators of the reactors and other leading experts:

- Has a commitment been made to convert this facility to LEU fuel?
  - If so, has it been documented?

- If a commitment has *not* been made to convert the facility, why not?
  - Where will the HEU come from to continue fueling the reactor?
  - What might encourage a commitment to convert to LEU?
- What is the fuel density required to convert the facility to LEU?
- Does the reactor core require structural modifications to convert the facility to LEU?

To explore progress on LEU fuel development, I posed the following questions to engineers engaged in the development activities, as well as to other leading experts:

- What are the anticipated densities for the UMo dispersion fuel and UMo monolithic fuel?
- What are the various chemical matrices being tried?
- What progress has been made in qualifying these LEU fuels?
- What challenges are being confronted in qualifying the LEU fuels?
- What is the current schedule of fuel testing?
- Which fuels are being developed for which reactors?
- For each reactor, what is the desired –
  - Neutron flux profile?
  - Fuel meat thickness?
  - Additional contour requirements for monolithic fuels ("foil profile")?
  - Fuel plate assembly?
- What other unique requirements exist for any of the reactors?

## Method

To answer these questions, I researched publications and open-source papers from recent scientific conferences, and I conducted interviews. The status and challenges of converting the HFIR reactor can be found in the latest publications by Trent Primm, who leads the conversion effort at the reactor. In addition, I met with Michael Whaley, who was a Senior Reactor Operator at HFIR and recalls the early debate about the prospects of converting the reactor. I also interviewed by phone Ronald A. Crone, Director of the HFIR's Research Reactor Division.

To research the ATR's conversion effort, I met with Josh Peterson, a visiting scientist at Idaho National Laboratory. He explained the computational modeling efforts used to verify and validate the new fuel being developed for that HPRR and others. He also highlighted the goals of conversion from an engineer's perspective.

For the three other U.S. HPRRs that still use HEU fuel, I conducted telephone interviews with officials at MURR, NBSR, and MITR. They explained the conversion status of the reactors, identified the fuel specifications needed, and noted any unique modifications required for conversion of their reactors. Specifically, I spoke to the following personnel:

- Les Foyto, Associate Director of Reactor and Facility Operations at MURR.
- Sean O'Kelly, Deputy Director at NBSR.
- Tom Newton, Associate Director, Reactor Engineering at MITR.

As I investigated the status of LEU fuel development, I learned that the monolithic fuel being developed at ATR was intended to enable conversion of all U.S. HPRRs, and that sample fuel plates were currently undergoing irradiation testing in the ATR. Test results are submitted to Argonne National Laboratory (ANL) for validation and verification using computational modeling. Accordingly, I spoke to officials who work for the U.S. Global Threat Reduction Initiative (GTRI) at Argonne, including John Holland, manager of the Engineering Application (fuels) Section of the RERTR program. He was very helpful in providing information about the fuel development efforts to convert both U.S. and European HPRRs. To further investigate developments in Europe, I contacted Dr. James E. Matos, manager of the HEU Minimization Methods Section of the RERTR program. He provided information on the progress and prospects of the LEONIDAS UMo dispersion fuel qualification program.

Colleagues on this book project, who traveled to Europe for their field research, also contributed to my overall findings concerning European research reactors. Kenneth Dayman's research documented the conversion prospects of Germany's FRM-II. Rebecca Ward investigated, on my behalf, the status of any conversion efforts for France's two HPRRs. But this chapter relies primarily on American sources, who were more forthcoming about the progress of trans-Atlantic LEU fuel development and conversion efforts.

## Findings: U.S. Reactors

Conversion can be costly. For example, HFIR officials determined in a feasibility study that it would cost $76 million to convert from HEU to LEU fuel. However, such costs can be and are being reduced somewhat by collaborative sharing of information between reactor operators, national laboratories, and fuel fabricators.[7] The GTRI program funds and leads such a collaboration between INL, ORNL, ANL, Los Alamos National Laboratory (LANL), the DOE's Y-12 uranium processing facility, and the private fuel fabricator Babcock & Wilcox (B&W) to qualify LEU fuel capable of converting all of the U.S. HPRRs.[8] They develop not only the basic LEU fuel design, but also the necessary technology to fabricate such LEU fuel on a commercial-scale at a cost comparable to HEU fuel assemblies.

It was this collaboration that developed the monolithic LEU. The fuel is a uranium-molybdenum alloy foil, composed of 10 percent molybdenum by weight and referred to as "U-10Mo." Full-sized fuel plates using U-10Mo have successfully undergone irradiation testing at INL.[9] The fuel was enclosed within a 6061 aluminum cladding using a zirconium bonding. This design provides a fuel-meat density of approximately 15.3 g/cc, sufficient to convert all five U.S. HPRRs.[10]

The next step will be to produce full-sized LEU fuel elements and irradiate them for testing. After that, the post-irradiation examination (PIE) will be used to create benchmark data, which for these reactors currently exists only for irradiated HEU fuel. ANL will then use the benchmark data from the LEU irradiation tests to enhance its computational modeling of LEU-fueled HPRRs. This will help assure the accuracy of the original computational modeling, used to predict the safety and performance of the reactors with LEU fuel.[11]

The Nuclear Regulatory Commission (NRC) licenses three of the U.S. HPRRs: MURR, NBSR, and MITR. They all have moderate power, operating at 10MW or less. To convert to LEU, these reactors are anticipated to use "base monolithic fuel," which has a relatively simple geometry. While the same U-10Mo fuel meat will be used for all the U.S. HPRRs, the geometry of the fuel plates is more complicated for the other two reactors operated by DOE.[12] (See Table 4.3 for the original conversion schedule, now delayed.)

MITR's commitment to convert is evident from its operator working closely with INL to develop fuel elements for irradiation testing. For MITR, the LEU plates will need to be thinner than the current HEU versions, so the number of plates will be increased to maintain the same fuel cycle. The power must also be increased from 6MW to 7MW to maintain the neutron flux. The reactor core will not require significant modifications, and the heat transfer capability will not be modified. The MITR reactor will undergo a gradual conversion, meaning that as spent HEU fuel elements are removed, they will be replaced with fresh LEU fuel until the entire core is eventually converted.

A prototype LEU fuel assembly for MITR is being fabricated for test irradiation in the ATR. The operator is expected to complete qualification of the fuel and submit a report to the NRC by late 2016. The Commission will then review the fuel qualification report, expected to take three months, and assuming it finds no concerns will issue a license for use of the fuel. The operator will then prepare the safety analysis, aiming to initiate conversion by early 2017.[13]

The NBSR is scheduled to convert by 2018. Feasibility assessments indicate that neither the core nor the power will need to be changed. The only anticipated modification will be an upgrade to the cold neutron source. The cold source at the NBSR uses liquid hydrogen to slow down the neutrons. With an upgrade of its cold source, the converted reactor is expected to have the "potential for small gains in cold neutron performance."[14] Research is also underway on methods that should prevent conversion from causing more than a 10 percent loss in neutron flux. Accordingly, the biggest perceived challenge is the scheduling of the fuel replacement, because the reactor typically operates 24 hours per day and is highly utilized.[15]

The MURR faces marginally more difficult, but still manageable, challenges to conversion. To maintain a comparable thermal neutron flux with LEU fuel, the reactor's power must increase from 10MW to 12MW. This raises the requirement for heat removal, so the operator will upgrade the cooling towers. The LEU fuel assemblies at this reactor also will have a reduced plate thickness.

Irradiation testing of a full-size LEU fuel element for MURR will be conducted in Belgium's BR-2 reactor.[16] Complete conversion of MURR to LEU fuel is

*Table 4.3* Original Conversion Schedule for the Three NRC-Licensed HPRRs

| *Schedule Event* | *MITR* | *MURR* | *NBSR* |
|---|---|---|---|
| Element Fabrication Complete | Jan 2014 | Jan 2014 | Dec 2013 |
| Irradiation Commencement | Feb 2014 | Apr 2014 | Feb 2014 |
| Irradiation Completion | Dec 2015 | Apr 2016 | Apr 2015 |
| Post-Irradiation Exam Complete | Oct 2016 | Jun 2017 | Feb 2016 |
| Conversion Request | July 2016 | July 2016 | Aug 2016 |

Source: N. E. Woolstenhulme and R. B. Nielson, "DDE Design Status Report Nov 2011," Idaho National Laboratory, November 2011, p. 11.

expected by 2019. Perhaps the greatest concern about the switch to LEU is the potentially extended down-time during conversion, because the reactor normally produces medical isotopes for hospitals (although not the most commonly used isotope, Mo-99). Eighty percent of the facility's budget reportedly is self-generated through isotope sales. Therefore, prolonged shutdown for conversion could create both a shortage of isotopes for hospitals and a financial shortfall for the reactor. Nevertheless, the operator is committed to convert to LEU fuel so long as it is safe to do so.[17]

The two other U.S. HPRRs – the HFIR and ATR – require a more complex fabrication process for the monolithic fuel because their fuel assemblies are composed of curved fuel plates. The ultra-high density of U-10Mo makes the material stiff and thus difficult to fabricate into such a curved shape. In addition, the HFIR requires a graded fuel meat within the plates, meaning that the base of the plate must be axially tapered and the rest of the plate must be radially tapered.[18]

At the HFIR, officials are developing new LEU fuel plate designs that will enable this conversion.[19] To maintain a similar neutron flux using the new LEU fuel, the HFIR is anticipated to operate at 100MW, instead of 85MW. The original design of the HEU core anticipated operation at 100MW, but the thermal hydraulic design of the new LEU core will need to be tested to ensure it can sufficiently cool the core during operation.[20] It is not known exactly when a full-sized LEU fuel element for the HIFR will be fabricated or test irradiated. But the HIFR is expected to be converted by 2020.[21]

The ATR will be converted last because much of the irradiation testing of the LEU fuel plates and elements for the U.S. HPRRs takes place at this facility. The reactor also requires the curved fuel plates that are a challenge to manufacture with the new U-10Mo fuel. A full-sized, monolithic ATR test plate was irradiated in the reactor in 2011.[22] It is not known exactly when a full-sized LEU fuel element for the ATR will be fabricated or test irradiated. But the ATR is expected to be converted by 2021.

## Findings: European Reactors

The LEONIDAS fuel qualification program is supported by the U.S. GTRI reactor conversion program and includes collaboration with two U.S. national laboratories, INL and ANL. It currently is conducting experiments to develop and qualify a high-density LEU fuel to convert two of the four European HPRRs: Belgium's BR2 and France's HFR. Unlike the U.S. two-track research program – which emphasizes monolithic fuel but also explores dispersion fuel – the European effort aims exclusively at a dispersion fuel made of U-7Mo in an Al-Si matrix. The uranium density is about 8 g/cc, using 19.7%-enriched uranium. The cladding for the fuel plates will be different for each reactor: AG3NE for the BR2, and AlFeNi for the HFR.[23] The lab scale production of such "E-FUTURE" plates has been completed, and irradiation tests under high flux conditions are currently underway at the BR2 reactor. To help ensure safety, SCK-CEN conducted flux and thermal-hydraulic computational modeling prior to this irradiation testing.

If the irradiation tests of the plates are successful, the next step will be the lab scale production of a complete fuel element for the BR2. LEONIDAS is not responsible for this step, because its mission ends with developing and qualifying the LEU fuel using the test plates. The reactor operators themselves are responsible for manufacturing fuel elements for testing using the qualified U-7Mo LEU fuel. The operators also have responsibility for any necessary modifications to their reactor cores to ensure safety and best performance of the fuel, and to preserve the experimental capabilities, fuel-cycle duration, and operating costs of their facilities.[24]

The operator of a third European HPRR, the Orphée reactor in France, has not made a commitment to convert. Although the fuel being developed by LEONIDAS has the potential to convert this reactor, Orphée officials reportedly say they are awaiting qualification and licensing of the U-7Mo LEU fuel – and the results of its performance in the HFR – before possibly studying the feasibility of converting their own reactor.[25]

Europe's final, HEU-fueled HPRR is the FRM-II in Germany. This reactor has a very small core that makes it more difficult to convert to LEU fuel, according to multiple feasibility studies. Dr. James Matos of the RERTR Program reports that it is not possible to convert the reactor using existing LEU fuels without enlarging the core. Even with the ultra-high-density fuels now under development, it probably would not be possible to convert the reactor to LEU without enlarging the core or significantly reducing performance. But if one of these new LEU fuels were successfully qualified, it should be possible to reduce the reactor's fuel enrichment to 50 or even 35 percent. So far, however, the reactor's operator has not committed to any reduction in fuel enrichment. For further details on the feasibility of converting the FRM-II, see the chapter in this volume by Kenneth Dayman.

For now, all four European HPRRs require HEU fuel to operate. In the past, the United States typically supplied HEU to all of these reactors, and for a very limited time in the 1990s Russia provided HEU to the FRM-II, HFR, and Orphée reactors. Under U.S. law (the so-called "Schumer Amendment"),[26] the United States can export HEU to fuel a reactor only if three conditions are met: the reactor cannot yet convert to LEU fuel; the operator has committed to convert as soon as possible;

and LEU fuel is being developed that would enable the reactor to convert. Currently, therefore, the United States is barred by law from exporting HEU to the FRM-II or Orphée. Dr. Matos believes that neither France nor Germany will go to Russia for HEU, although both have done so once in the past. Bertrand Barré of AREVA argues that because science is a competitive business, until the United States converts the HFIR, foreign HPRRs will not want to put themselves at a competitive disadvantage by converting to LEU fuel.[27]

## Summary and Recommendations

All five U.S. HPRRs have made a commitment to convert to LEU monolithic fuel. A collaborative effort that includes national laboratories has helped to reduce fuel development costs and to facilitate conversion by sharing LEU fuel benchmarking data. In Europe, two of the four HPRRs – the BR-2 and HFR – have committed to convert, and the LEONIDAS partnership is currently irradiation testing an LEU dispersion fuel to facilitate their conversion. Thus, of the nine HPRRs in the United States and Europe, seven are on track to convert to LEU fuel by 2021.

The remaining two HPRRs in Europe – France's Orphée and Germany's FRM-II – have not committed to convert to LEU fuel. The United States can and should take several steps to encourage the operators of these facilities to convert. Coercive leverage arises from the fact that both reactors eventually – the FRM-II more urgently – will require fresh HEU fuel, which they apparently are unwilling or unable to acquire from Russia. The United States should strictly enforce the Schumer Amendment. Prior to qualification of the ultra-high-density LEU fuel, this would permit U.S. export of HEU to fuel the Orphée reactor only if its operator committed to convert to LEU fuel as soon as possible. One essential indicator of such a commitment, in addition to a written pledge, would be the operator initiating a conversion feasibility study.

United States law does not permit export of HEU to the FRM-II because no fuel is under development that would enable the reactor to convert to LEU. But if that facility's operator were willing to reduce the fuel's enrichment to 35 percent or below, the United States could consider facilitating the supply of such uranium from existing U.S.-origin stocks in Europe. Meanwhile, EU officials should discourage the operators of these two recalcitrant facilities, and their national regulators, from seeking HEU from Russia. The United States also can help discourage Russian exports of HEU by continuing to work with Russia on converting its own research reactors to LEU fuel.[28]

Finally, the United States should use its moral leverage. Washington has expended significant resources and taken concrete steps to phase out HEU fuel from its own HPRRs. It should use this moral high ground to persuade the operators of the two European HPRRs that have not yet committed to convert to LEU fuel that such conversion is feasible, safe, cost-effective, and vital for global security.

## Notes

1 John G. Stevens, "Technical Challenges for Conversion of U.S. High-Performance Research Reactors (USHPRR)" presentation, Argonne National Laboratory, 29 November 2010, slide 8.

2 Daniel M. Wachs, Curtis R. Clark, and Randall J. Dunavant, "Conceptual Process Description for the Manufacture of Low-Enriched Uranium-Molybdenum Fuel," U.S. Department of Energy, Office of Nuclear Nonproliferation, Idaho Falls, February 2008, p. 4.

3 Alan J. Kuperman, "Appendix 2," in *Nuclear Power and the Spread of Weapons: Can We Have One Without the Other?*, eds. Paul L. Leventhal, Sharon Tanzer, and Steven Dolley (Washington, D.C.: Brassey's, 2002), p. 254.

4 Alexander Glaser and Frank von Hippel, "On the Importance of Ending the Use of HEU in the Nuclear Fuel Cycle: An Updated Assessment," paper presented to the RERTR Meeting, San Carlos de Bariloche, Argentina, 3–8 November 2002, p. 13.

5 CERCA, the Company for the Study of Atomic Fuel Creation in France, is an AREVA subsidiary that develops and fabricates nuclear fuel. CEA, the Commissariat à l'énergie atomique et aux énergies alternatives, is a technical research organization funded by the French government. SCK-CEN, Studiecentrum voor Kernenergie – Centre d'Etude de l'énergie Nucléaire, is a nuclear research organization in Belgium. ILL, Institut Laue-Langevin, is a French neutron science and technology research center.

6 A. Bergeron, A. Tentner, and J. G. Stevens, "Feasibility Analyses for HEU to LEU Fuel Conversion of the Laue Langevin Institute (ILL) High Flux Reactor (RHF)," Nuclear Engineering Division, Argonne National Laboratory, 2010, p. 8.

7 Tracey Guida and R. T. Primm, III, "Establishing a Cost Basis for Converting the High Flux Isotope Reactor from High Enriched to Low Enriched Uranium Fuel," Research Reactor Division, Oak Ridge National Laboratory, 2010, p. 14.

8 Telephone interview with John Holland, section manager of Engineering Application (fuels) Section, RERTR on 18 March 2011.

9 A. B. Robinson, D. M. Wachs, G. S. Chang, and M. A. Lillo, "Summary of 'AFIP' Full Sized Plate Irradiations in the Advanced Test Reactor," Idaho National Laboratory, 2010, p. 6.

10 G. F. Vandegrift, J. Jerden, D. C. Stepinski, J. Figueroa, M. A. Williamson, M. A. Van Kleeck, R. J. Baskovitz, A. J. Ziegler, L. E. Maggos, J. Swanson, J. Fortner, and A. J. Bakel, "Reprocessing of LEU U-Mo Dispersion and Monolithic Fuels," paper presented to the European Nuclear Society European Research Reactor Conference, Rome, Italy, 20–24 March 2011, p. 1.

11 Telephone interview with Tom Newton, Associate Director, Reactor Engineering at MITR, 11 March 2011.

12 Telephone interview with John Holland, RERTR manager of Engineering Application (fuels) Section, 18 March 2011.

13 Telephone interview with Tom Newton, 11 March 2011. N.E. Woolstenhulme and R.B. Nielson, "DDE Design Status Report Nov 2011," Idaho National Laboratory, November 2011, http://www.inl.gov/technicalpublications/Documents/5223060.pdf (accessed on 27 September 2012). Personal communication with Alan J. Kuperman, 12 May 2012.

14 John G. Stevens, "Technical Challenges for Conversion of U.S. High-Performance Research Reactors (USHPRR)," presentation, Argonne National Laboratory, 29 November 2010, slide 15.

15 Telephone interview with Sean O'Kelly, Chief of Reactor Operations, NBSR, 16 March 2011.

16 Woolstenhulme and Nielson, "DDE Design Status Report."

17 Telephone interview with Les Foyto, Associate Director of Reactor and Facility Operations, MURR, 16 March 2011.

18 Telephone interview with Ronald A. Crone, Director of Research Reactor Division, HFIR, 1 April 2011.

19 R.T. Primm III and Germina Ilas, "Reference (Axially Graded) Low Enriched Uranium Fuel Design for the High Flux Isotope Reactor (HFIR)," paper presented to the RERTR meeting, Knoxville, Tenn., 19–23 September 2010, pp. 3, 8.

20 R. T. Primm III, D. H. Cook, J. D. Freels, G. Ilas, B. C. Jolly, J. H. Miller, D. L. Pinkston, D. G. Renfro, and J. D. Sease, "Design Study for a Low-Enriched Uranium Core for the High Flux Isotope Reactor, Annual Report for FY 2010," Oak Ridge National Laboratory, February 2011, pp. 3, 13.

21 Frank Munger, "Eight more years of HEU fuel at HFIR," *Atomic City Underground*, March 2012, http://blogs.knoxnews.com/munger/2012/03/eight-more-years-of-heu-fuel-a.html (accessed on 27 September 2012). See also, D. G. Renfro, et al., "Low-Enriched Uranium Fuel Conversion Activities for the High Flux Isotope Reactor, Annual Report for FY 2011," ORNL/TM-2011/507, Oak Ridge National Laboratory, March 2012, http://info.ornl.gov/sites/publications/files/Pub33904.pdf (accessed on 27 September 2012).

22 Glenn Moore and M. Craig Marshall, "AFIP-6 Fabrication Summary Report," INL/EXT-11-21655, September 2011, Idaho National Laboratory, http://www.inl.gov/technicalpublications/Documents/5201315.pdf (accessed on 27 September 2012).

23 Telephone interview with John Holland, 18 March 2011.

24 F. Frery, E. Koonen, P. Lemoine, and C. Jarousse, "LEONIDAS U(Mo) Dispersion Fuel Qualification Program: Progress and Prospects," paper presented to the RERTR Meeting, Lisbon, Portugal, 10–14 October 2010, pp. 1–10.

25 Telephone interview with James Matos, RERTR Program, HEU Minimization Methods Section, 18 March 2011.

26 § 134., 42 U.S.C. § 2160d, commonly known as the Schumer Amendment to the U.S. Energy Policy Act of 1992, Pub. L. No. 102–486, 106 Stat. 2776 (1992).

27 Interview by Rebecca Ward of Bertrand Barré, Scientific Advisor to AREVA, 2 February 2011.

28 Telephone interview with James Matos, 18 March 2011.

# 5 Russia

## Icebreaker Ships and Floating Reactors

*Christine Egnatuk*

Russia is the only country in the world that currently operates civilian, nuclear-powered ships. Its fleet mainly comprises nuclear-propelled icebreakers that each winter clear the "Northern Route" shipping lane from the Atlantic to the Pacific along the Russian Arctic coast, breaking paths through the ice from Murmansk to the Far East.[1] The reactors that propel Russia's fleet use fuel of highly enriched uranium (HEU), a nuclear weapons-usable material that poses security risks and sets a dangerous precedent for other states. Accordingly, this chapter assesses the technical and political feasibility of converting the propulsion reactors in Russia's nuclear icebreakers from fuel of HEU to low enriched uranium (LEU), which is unsuitable for nuclear weapons. As part of this analysis, the chapter compares the existing HEU-fueled icebreaker reactors to Russia's closely related, prototype Floating Nuclear Power Plant (FNPP), fueled with LEU.

## Introduction

The Russian nuclear icebreaker fleet is the focus of international attention because of two major concerns. The first is the environmental impact on the region from the icebreaker fleet and Russia's nuclear navy. The second concern, which motivates this research, is the security risk associated with the regional storage of fresh and spent nuclear fuel, both of which have presented problems in the past. Despite some recent improvements in physical security, the long-term effectiveness of these upgrades is still to be determined.

The world's largest concentration of HEU is stored on the Kola Peninsula in northern Russia. The HEU includes spent and fresh fuel for both the Russian nuclear icebreaker fleet and the Russian nuclear submarine fleet. In 1993, there were two reported thefts of fresh HEU fuel from the region.[2] Both assemblies were recovered, but the fuel was missing for several months before the thieves were caught.

Improper disposal of spent fuel raises both environmental and security issues. Russia's Northern Fleet operates about two-thirds of the country's nuclear propulsion vessels, both civilian and military. Spent nuclear fuel is not properly secured,[3] and some has been dumped in the Barents and Kara seas with no physical protection.[4] There are approximately 100 submarine fuel cores (21,000 fuel assemblies) in the Andreeva Bay.[5] There is an estimated 32 tons of HEU spent fuel within the over 100 tons of spent fuel. The cost for a potential clean-up was estimated at $4 billion in 2006.

In September 1999, security systems were upgraded on the *Imandra* refueling ship,[6] and at the Atomflot port (see Figure 5.1) that houses the icebreaker fleet, funded by the U.S. Material Protection, Control and Accounting (MPC&A) program.[7] The Murmansk Shipping Company, operated by the Ministry of Transportation, was the only commercial shipping facility upgraded. The upgrades included only the protection of fresh nuclear fuel, which is currently HEU. Some of the spent nuclear fuel may also have enrichment above 20 percent, which would put it in the classification of HEU. The enrichment levels of the spent nuclear fuel would depend on the original enrichment of the fuel and the discharge burn-up. Ten years after removal from the reactor, the fuel is no longer "self-protecting." That is, the spent fuel has been allowed to decay, or cool, long enough that if nefarious actors stole it, they could transport the fuel with minimal shielding and still avoid a lethal radiation dose.

The current operational icebreakers, all of which are propelled by reactors using HEU fuel, are approaching their service lifetimes. The next generation of icebreakers is expected to employ a new reactor, the RITM-200, which is being designed to use LEU. If this planned transition to a new generation of icebreakers were on schedule, Russia would soon end its use of HEU for civilian propulsion. Unfortunately, Russia now plans to extend the ship and reactor service life of the current fleet in order to save money, thereby perpetuating the use of HEU fuel. Several research groups accordingly have explored the feasibility of converting one of Russia's *current* civilian propulsion reactors, the KLT-40, from HEU to

*Figure 5.1* Atomflot facilities. The left picture is the floating dock that Atomflot obtained in 1995. This dock is used to maintain part of the nuclear-powered icebreaker fleet (including Taymyr and Vaygach) and service ships. The right picture is the Atomflot enterprise. The activity at this port includes servicing the icebreakers and service ships.

Source: Bellona (left) and Rosatom Flot.

LEU fuel. As elaborated below, the KLT-40 indeed could be fueled with LEU without any modifications to the reactor vessel, but there would be an economic penalty for enrichment below 30 percent.

The next section of this chapter reviews, and identifies shortcomings within, the existing literature on the topic. The lack of information on the KLT-40 reactor design, and conflicting reports about its fuel enrichment level, are two of the biggest problems. The methodology section includes my research questions and the officials whom I interviewed. The findings section discusses the next generation icebreaker propulsion reactor and analyzes the plans for conversion of existing KLT-40 propulsion reactors. I also compare the KLT-40 to the KLT-40S, which is employed in the floating nuclear power plant, and explain how the similarities allow for the use of the same LEU core assembly. The chapter also proposes an economic feasibility study of a centralized reactor design for all propulsion and small modular reactors produced by Russia. My conclusion explores the counter-intuitive possibility that converting Russia's existing icebreaker fleet to LEU fuel might actually be economically preferable to the continued use of HEU.

## Literature Review

One of the main concerns with the use of HEU fuel for the icebreaker fleet is the security of the nuclear fuel storage. Two fuel assemblies (a total of 3.6 kg of 36%-enriched HEU) were stolen from a naval fuel store on the Kola Peninsula in July 1993 (see Figure 5.2).[8] Another theft occurred in the same region only a few months later, in November 1993, when 4.5 kg of 20%-enriched fuel was stolen from Sevmorput shipyard in Murmansk by two former naval officers. As noted, the fresh fuel storage facilities in the region have been enhanced by U.S. MPC&A funds. But such upgrades are not nearly as effective against the threat that insiders – i.e., facility employees – might attempt to steal HEU.[9] The U.S. program also did not upgrade security for spent fuel, much of which is no longer self-protecting, as noted.

Russia's civilian nuclear propulsion program was an offshoot of its nuclear submarine program. The civilian nuclear fleet currently comprises both icebreakers and cargo vessels. The fleet reflects decades of development and is currently operating three different reactor types: the OK-900A, KLT-40M, and KLT-40.[10] Russian officials report that the nuclear icebreaker fleet is fueled with HEU. The Murmansk Shipping Company, a private entity that formerly operated and currently maintains the icebreaker fleet, claims the fuel enrichment is between 30–40 percent.[11]

However, in 1990, the Norwegian government obtained contradictory information in a safety document from a distressed Russian nuclear icebreaker, the *Sevmorput*, which needed to enter a Norwegian port. The safety report indicated that the reactor contained 150.7 kg of U-Zr fuel with 90 percent enrichment. The thermal power output was reported as 135 $MW_t$,[12] and the uranium density of the fuel was 4.6 $gU/cm^3$.[13] Using the information from the safety report, a group of Norwegian scientists explored the feasibility of conversion

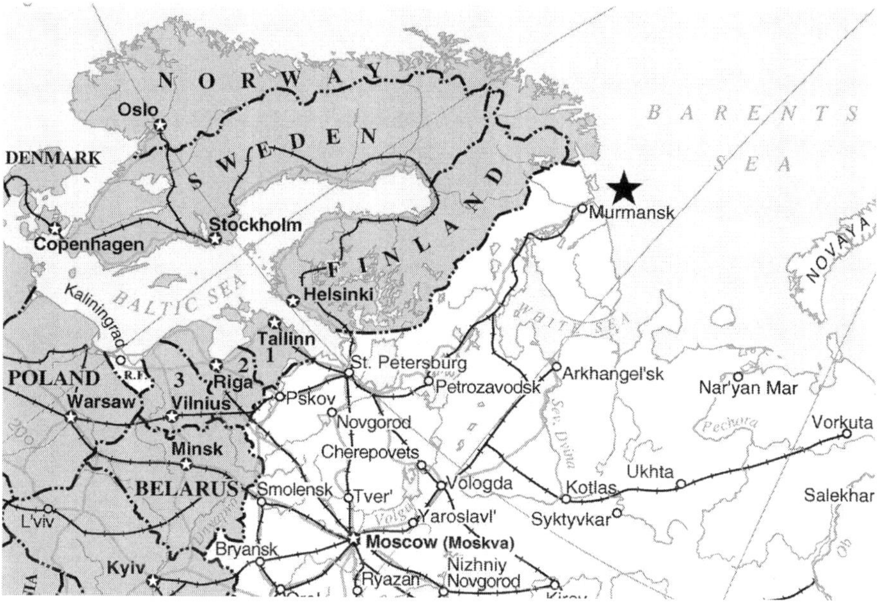


*Figure 5.2* Murmansk, located on the Kola Pennisula, is marked by the black star on the map.

Source: United Nations.

to LEU.[14] Using a model that assumed hollow HEU fuel rods in the existing reactor, they determined that solid LEU fuel rods would provide a similar thermal power output. Using the same reactor core design proposed by the Norwegian scientists, a group of Russian scientists considered varying the fuel type – to either $UO_2$ in an aluminum matrix, or a uranium alloy with aluminum – exploring enrichment levels of 20 and 40 percent.[15] Both the Russian and Norwegian studies agreed that a thermal output close to the value in the safety report was obtainable with LEU fuel.

The core design of the existing reactor in the previously mentioned models is only an approximation, inferred from the safety report. But if the fuel volume actually could be held essentially constant while using LEU instead, that would facilitate conversion. This could be enabled by higher density LEU fuel that has the same amount of U-235 per unit volume as traditional, lower uranium density HEU fuel. Ultra-high density LEU fuels are being developed in several countries for use in research reactors by Russia, the United States, and Europe in conjunction with the Reduced Enrichment for Research and Test Reactors (RERTR) program.[16] Very high density (6.0 $gU/cm^3$) uranium-molybdenum fuel has already been developed in Russia and achieved a high burn-up (an average of 83.8 percent of the original U-235 was burned up) in preliminary tests on miniature elements.[17]

The current fleet of icebreakers is aging, and the new generation of icebreakers is expected to be powered by a new reactor, the RITM-200, designed by OKBM Afrikantov to use LEU fuel. The design achieves a smaller size (1,100 tons) by placing the steam generator inside the reactor vessel. The size of the vessel is 6m x 13.2m x 15.5m. The integral reactor design also decreases the material costs of construction by enabling shorter pipes. The drawback is that the design may increase the time of routine maintenance. Many of the valves and other fittings will be within the reactor vessel, requiring the technician to disassemble a large part of the reactor to fix or replace something that in the current design is outside of the vessel. The anticipated increase in downtime due to maintenance would necessitate more ships to produce the same amount of work, holding everything else constant.

But the OKBM designers claim the reactor has a flatter neutron flux profile that will reduce maintenance needs.[18] The flatter flux profile is probably achieved with varying radial enrichment zones in the reactor core. This variation produces a lower flux at the center of the reactor, reducing the power density and heat. Such lower heat and radiation levels at the center of the core should reduce deterioration of the instrumentation there, thereby reducing maintenance needs. This could compensate, to some degree, for the greater difficulty of conducting maintenance on a reactor with an integral design.

The RITM-200 features two identical reactors per ship, producing a combined thermal power of 175$MW_t$. The reactor facility output is expected to be about 65 percent capacity, which accounts for both reactors operating simultaneously about 30 percent of the time. Assuming 65 percent capacity, the reactors would need to be refueled every seven years. The reactors and ship are also planned to be refurbished once, after 20 years, to yield an expected service life of 40 years. Although there are no published safety records for past, current, or future icebreaker reactor designs, the RITM-200 development group claims the new reactor is safer than its predecessors. The designers also claim that the new reactor is competitive with non-nuclear power plants and can be used for ships, floating power plants, or indeed anywhere a stable source of power is needed.

The prototype floating reactor, the KLT-40S, is essentially an LEU version of the KLT-40, which is an HEU-fueled reactor used in the latest icebreakers. Russia launched the hull of the first floating plant, the *Akademik Lomonosov*, in July 2010,[19] opting for the LEU design in response to international concerns. In 2000, a partnership was first proposed between the U.S. Argonne National Laboratory and Russia's Bochvar Institute to develop high density LEU dispersion fuel for the floating nuclear power plant designs.[20] The project fell through, but the KLT-40S reactor plant ultimately was designed to use LEU dispersion fuel of 13–16 percent enrichment.[21] The KLT-40S requires refueling every two to three years,[22] on-site fuel storage,[23] and on-site refueling, which entails removing spent fuel from the reactor and replacing it with fresh fuel while the barge on which it is mounted remains docked at the city where it produces power.[24] An alternative reactor design, the KLT-20, is more appealing because it has a longer, 10-year core life,[25] enabling refueling to be conducted offsite, while still using LEU fuel (19.2 percent enriched).[26] The KLT-20 was developed as the second generation of barge-mounted reactor, but there are no immediate plans to implement it on such a floating power plant.

The actual design of the KLT-40 core is not published. Nor does public information adequately explain the differences between the KLT-40S (the LEU reactor used in floating power plants) and the KLT-40 (the HEU reactor used in the icebreaker fleet). Therefore, it was necessary to question experts in the area of icebreaker propulsion and small modular reactor designs to elucidate these differences.

## Methodology

My research sought to establish the current status of all ships within the Russian nuclear icebreaker fleet managed by the Murmansk Shipping Company under Russia's Department of Transportation and Rosatom. I also sought to verify the enrichment of the fuel used in the propulsion reactors – now and in the past. Most importantly, I explored the technical and political prospects and challenges of converting the current icebreaker reactors to LEU using Russian developed fuels, including the possibility of support from a third party. As one potential way to lower the cost of conversion to LEU, I also explored the option of gradually inserting LEU fuel rods into the core throughout the refueling schedule as HEU fuel rods were removed.

The future plans for the icebreaker fleet, including the RITM-200 reactor type, needed to be substantiated, as did the alleged benefits of the proposed reactor over the current design. My research also sought to obtain the safety analysis of the new reactor design to assess the possible environmental hazards and to carry out a full probabilistic risk assessment.

The first phase of the project comprised a thorough literature search of published papers, global conferences, and web articles. The second phase included contacting experts in the field through email and by phone. The third and final phase of the research included face-to-face interviews, when possible, or otherwise Skype conversations.

The following people (listed in alphabetical order) were interviewed during the course of this research project:

- Nikolay Arkhangelskiy, Advisor for the Directorate for Scientific and Technical Complex within Rosatom.
- Igor Kudrik from the Oslo, Norway office of Bellona, an NGO focused on environmental issues, which also has offices in Brussels, Washington, DC, and Murmansk and St. Petersburg in Russia.
- Vladimir Kuznetsov, formerly of Rosatom's Department of International Cooperation, and currently working for the IAEA on small modular reactor design technology.
- Alexander Nikitin, Chairman of ERC Bellona (St. Petersburg).[27]
- Ole Reistad from the Norwegian Radiation Protection Authority.
- Yevgeny Sidorov, Chief Specialist at the State Scientific Center of the Russian Federation of Physics & Power Engineering.
- Alexander Zhukov, Engineer at the State Scientific Center of the Russian Federation of Physics & Power Engineering.
- Andrey Zolotkov, Head of Bellona-Murmansk.[28]

The major obstacle in this research project was finding people who were knowledgeable and willing to be interviewed about the nuclear icebreaker fleet. Many refused to be interviewed, did not respond to email messages or phone calls, or claimed to have no information about the nuclear fleet even though past publications and job positions indicated otherwise. There were also differing opinions on the future of the fleet. The next section documents all of the views expressed in the interviews, even though the validity of some of the statements may be questionable, as discussed in my subsequent commentary.

## Findings

The future of the nuclear icebreaker fleet is seen as promising among the Russian officials interviewed. The fleet is now under the control of Rosatom, which feels it is improving the management and in turn the economics. The fleet is serviced at the Atomflot base about 2 km from Murmansk. This service base is under control of the Ministry of Transportation. Atomflot is the location of the reactor refueling, maintenance on the vessels and reactors, receipt of fresh fuel, and storage of spent nuclear fuel (and other solid and liquid radioactive waste products).

In order to keep the Northern Route open and accommodate foreign commercial ships, Russian experts say that only five to seven nuclear icebreakers are necessary.[29] The diesel icebreaker fleet will also remain functional but does not have the range necessary to keep the Northern Route open year-round. Currently the diesel fleet is used for smaller and shorter trips and is not located at the same port as the nuclear icebreaker fleet.

### *Next Generation Propulsion Reactor*

The proposed RITM-200 reactor design (see Figure 5.3) is more efficient and economical than its predecessors.[30] It is expected to be used in the next generation icebreaker. Two to three new icebreakers are anticipated to be built in the near future, with a total of four new icebreakers ultimately planned to use the new RITM-200 reactor.[31] The designers claim a decreased power density in the core and an increased power capacity: 175 $MW_t$ compared to the KLT-40's 150 $MW_t$.

The RITM-200 reactor will be fueled with uranium enriched to less than 20 percent.[32] However, the designers claim the economies would improve if fuel enrichment were increased to 30–35 percent.[33] The reactor was designed to use LEU fuel in order to facilitate possible export for barge-mounted or other small modular usage. One Russian official explained the choice of LEU as "political correctness – keeping nuclear installations of civil destination (e.g. icebreakers) in line with the IAEA requirements and recommendations."[34] A member of Bellona said the proposed shift to LEU fuel is "purely to meet IAEA recommendation for enrichment for commercial reactors."[35] He also expressed doubt that the reactor would actually use LEU.

*Figure 5.3* The RITM-200 Reactor by OKBM. The integral design places the steam generator inside the reactor vessel. The flattened neutron flux profile is probably due to radial zones of varying enrichment.

Source: Bellona.

### *Current Fleet*

There are currently five operational nuclear icebreakers: *Rossia, Vaygach, Yamal, 50 let pobedy,* and *Taymyr.*[36] Two additional civilian, nuclear-powered ships are temporarily out of operation: the icebreaker *Sovetskiy Souz* and the light cargo carrier *Sevmorput*. The remaining ships – *Arctica, Sibir,* and *Lenin* – are being, or have already been, decommissioned. The *Lenin* was decommissioned in 1989 and subsequently converted into a museum in Murmansk Harbor, at an estimated cost of $1 million.[37]

As noted, Russia's icebreaking ships are reaching or have exceeded their originally planned service lifetimes. But in response to economic troubles, these service lifetimes of the ships, as well as of their reactors, have been extended. The original ship service lifetime was expected to be 100,000 hours. But *Arctica* had a life of 175,000 hours before being decommissioned in 1993. Only *Yamal* and *50 let pobedy* have yet to reach their original service lifetime of 100,000 hours. The planned service lifetime extension to 175,000 hours would add another five to seven years for each ship. This would enable *Taymyr* and *Rossia* to operate until 2018,[38] *Sovetskiy Souz* and *Vaygach* until 2020, and *Yamal* and *50 let pobedy* even longer.

The HEU currently being used in the nuclear icebreakers is a $UO_2$ dispersion fuel in an inert matrix. The fuel is currently fabricated in JSC "Mashinostroitelny Zavod," which is part of the JSC "Tvel" of Rosatom.[39] After fabrication, the fuel assemblies are shipped to the Atomflot base in Murmansk. The Russians claim this fuel design is proliferation resistant because the spent fuel contains little plutonium.[40] But this ignores the high U-235 content in the fresh and spent fuel, which obviously does raise proliferation and terrorism concerns.

The reactor core of the nuclear icebreaker fleet is referred to as a cassette design in the literature, which does not fully elucidate the concept. The cassette design allows for the removal of the entire core at one time, enabling quick refueling. This design also accommodates different fuel arrangements and varying enrichment of the fuel in the reactor. Therefore, the enrichment and core design of a propulsion reactor can be changed easily. For example, the projected range of fuel enrichment for use in the KLT-40 type reactor design is 15 percent to 90 percent.[41] Nikitin claims the icebreaker fleet reactors have served as a testing ground for core designs used in the submarine fleet and for the barge-mounted reactors – the latter of which would imply that LEU fuel may already have been used in an icebreaker reactor.

### *Future Plans for the Icebreaker Fleet*

As of late 2010, the scientists at OKBM expected that the majority of the current nuclear icebreakers would be decommissioned by 2016, but this is doubtful, as explained below.[42] In 2017, Russia plans to put into operation the first new icebreaker using the new RITM-200 reactor, with two more such vessels projected by 2027. A study in late 2010 estimated that the construction costs of the new icebreaker design using the RITM-200 had increased considerably. The current cost estimate for the first new icebreaker is $1.1 billion, and Russia's Duma has approved $630.5 million for the project through 2014.[43]

*Table 5.1* Current and Proposed Reactor Designs with their Designated Fuel Types

| *Reactor Design* | *Purpose* | *Fuel Enrichment* |
|---|---|---|
| KLT-40 | Icebreaker Propulsion (current design)* | 30%–40%, 90% |
| KLT-40S | Floating power plant (prototype design)** | <20% |
| RITM-200 | Icebreaker Propulsion (proposed design) | <20% |

*The fuel has enrichment of 30–40% according to the Murmansk Shipping Company, but 90% was reported to the Norwegian Port Authority. Diakov et al., "Feasibility of Converting," 34.

**The fuel will probably have varying enrichment of 13.1 to 18% within the core because of the use of radial enrichment zones for a flatter flux profile. Kuznetsov interview.

The lifetime extension of an old icebreaker from 100,000 hours to 175,000 hours costs about 20 percent of the price to build a new icebreaker.[44] Since the nuclear icebreaker fleet is struggling financially,[45] the service life extensions will probably be carried out on all applicable vessels in the fleet. Therefore, the reactor assemblies on the existing fleet could be fueled with HEU at least until *Sovetskiy Souz* and *Vaygach* reach their service lifetimes around 2020. A Rosatom official stated that the nuclear icebreaker fleet would be converted to LEU in the future, but he was unable to offer any additional information.[46] This same question was posed to a Russian scientist.[47] He claimed that conversion of existing reactors to LEU is theoretically possible, but would require additional justification and research, and the funding to carry out that analysis.

### *Comparison of the KLT-40 and KLT-40S reactor design*

The reactor in the most recent generation of operational icebreakers is the KLT-40 or KLT-40M (which just means the modified KLT-40). The KLT-40S, as shown in Figure 5.4, is the reactor design in the prototype floating plant. This FNPP design is basically the same reactor as the KLT-40 but uses LEU fuel and includes passive safety features required for commercial power production.[48] The FNPP design also includes additional equipment for electricity and steam production.[49]

When the Russians announced their plans to create a barge-mounted reactor plant using the same reactor design as their nuclear icebreaker fleet, there was cause for alarm. It was feared that the HEU fuel in the reactor vessel and a substantial amount of fresh fuel for refueling would be exported to various parts of the globe, including non-nuclear weapons states. In response, a Russia–U.S. partnership was proposed to develop high-density LEU fuel to enable conversion of the KLT-40. The A.A. Bochvar All-Russia Research Institute of Inorganic Materials (VNIINM) and The Nizhny-Novgorod Machinery Construction Plant (OKBM) in Russia were to partner with the U.S. Argonne National Laboratory to create a core design that could use high density LEU. Unfortunately, the collaboration fell apart before the U.S. side commenced work on it.[50]

Nevertheless, due to the flexibility of the core arrangement offered by the cassette design, the Russians were able to produce a design for the KLT-40S that is fueled with LEU dispersion fuel. The switch from HEU to LEU fuel was made in order to ease road-blocks to the export of the FNPP.[51] In order to comply with restrictions of the Nuclear Suppliers Group, Russia plans to lease rather than sell the plants to other countries.[52] The proposal is to sell the power to the country, but to keep the FNPP under Russian control and operation. Russia has evaluated the environmental ramification of the FNPP under normal conditions,[53] but a comprehensive study of the risk to the population and environment from sabotage or a natural disaster scenario has yet to be considered.[54]

The FNPP houses two reactors, each with 150MW$_t$ capacity. The total output of the reactor plant is 300 MW$_t$ with both reactors operating. The plant has 70MW$_e$ net power. There are some proposed designs that harness some of the heat to fuel

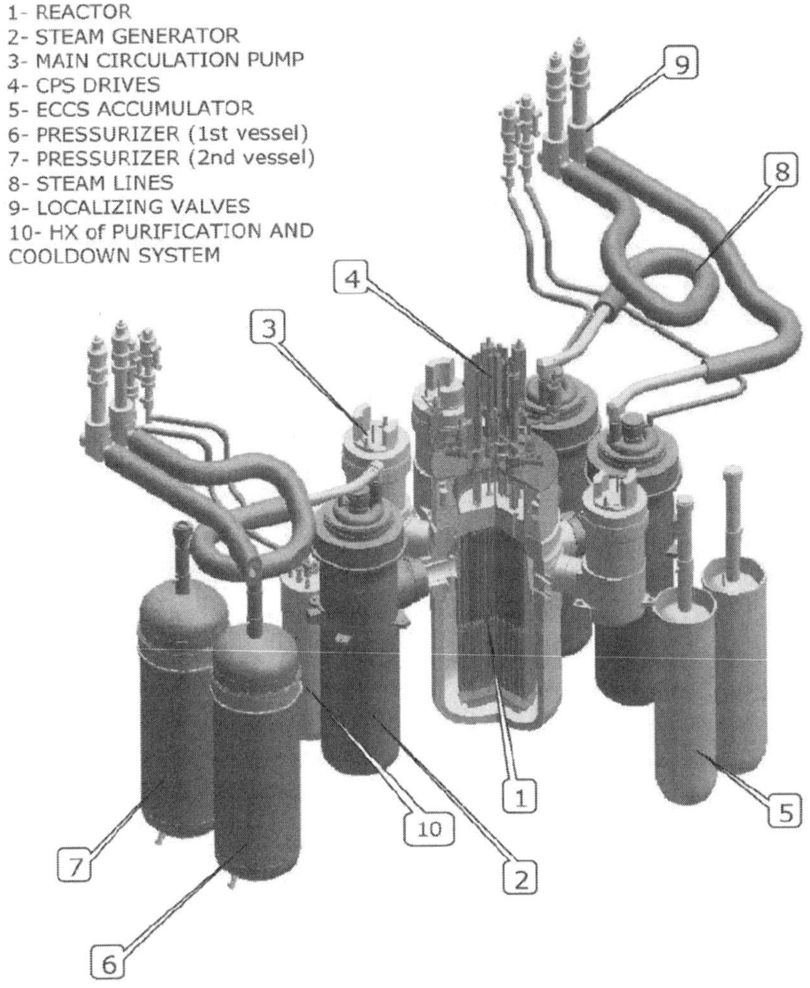


*Figure 5.4* KLT-40S reactor design. The system has four-loops and is cooled with light water. The steam generator is a once-through set-up with an external gas pressurizer system.

Source: "Design Features to Achieve Defence in Depth in Small and Medium Sized Reactors," IAEA Nuclear Energy Series, No. NP-T-2.2 (2009), 86.

an attached desalination plant designed in Canada. The KLT-40S has an 8,000 hour continuous operation period and a 40-year service lifetime. The average burn-up of discharged fuel is 45.5 MWd/kg. The initial fuel enrichment is still being decided but will be under 20 percent. Andrey Zolotkov claims the fuel is $UO_2$ in a silumin matrix with 18.6 percent enrichment.[55] By contrast, Vladimir Kuznetsov stated that there will be two radial enrichment zones within the reactor core, enriched to 15.7 percent and 13 percent.[56]

The prototype FNPP is being built in Baltiyskiy shipyard in St. Petersburg. The hull of the FNPP was set afloat in July 2010. After construction and testing is complete, the floating plant will be moved to Viluchinks in the Kamchatka region.[57] Three other cities are proposed to receive an FNPP: Pevek and Cherskiy, in the Chukotka autonomous region, and Ust-Kuya in the Sakha Republic. The estimated cost of constructing an FNPP, as of March 2010, was 15 billion rubles, which is more than $500 million.

In light of this high price for the FNPP, and the rising cost estimate of the next generation icebreaker, I raised the idea with Russian officials of creating a unified reactor design for all such projects that utilize a small thermal output. This would include the nuclear icebreaker fleet, nuclear submarine fleet, and the FNPP. Based on economies of scale, such a consolidation could help reduce the price of each reactor, and this benefit could extend to the fuel fabrication process. One Rosatom official said it was an interesting idea that would need much more research.[58] Another Rosatom official argued that the number of reactors to be built was insufficient to benefit from economies of scale.[59] A scientist at Obninsk offered to help find the scientists necessary to carry out the economic analysis if funding were provided.[60]

### *Safety of the Reactor Designs*

The KLT-40 reactor safety reports are not published. This is troubling to many researchers in the region because the lack of reporting on past accidents means that they know very little about the hazards of this reactor design.[61] Based on past experience in the region, this concern about environmental hazards is valid. But when the safety issue was posed to a Rosatom official,[62] he claimed the design was safe. He cited the example of the reactor in a nuclear submarine that sank to the bottom of the Northern Sea. The vessel was retrieved after two years and the reactor reportedly was able to be restarted. He also claimed that the environmental impact of the current KLT-40 reactor was minimal because it used less fuel and had lower temperature and pressure than previous designs. The safety analysis for the proposed reactor design, RITM-200, was also unavailable.

### *Spent Nuclear Fuel*

All of the spent nuclear fuel from the nuclear icebreaker fleet is currently being stored on service ships: the *Lepse, Lota* and *Imandra*.[63] The fuel has either stainless steel or zirconium cladding. Only the fuel that has stainless steel cladding can be reprocessed. After the mandatory cooling period, the spent nuclear fuel with stainless steel cladding can be shipped to a reprocessing plant. The spent zirconium-clad nuclear fuel that cannot be reprocessed will be moved to a land-based storage facility near the Atomflot base.

The exact amount of spent nuclear fuel on storage ships or stored on land near Atomflot is unknown. None of the experts interviewed claimed to know the amount of the spent nuclear fuel discharged by the icebreaker fleet or its enrichment levels. But Bellona reports that, as of 2001, there are 1,290 spent fuel assemblies on *Imandra* and 639 spent fuel assemblies on *Lepse*.[64]

Considering the varying initial enrichment levels and differing burn-up levels, there are probably very few, if any, people who can accurately categorize the waste from the icebreaker fleet. The fuel is assumed to be self-protecting for 10 years after discharge from the reactor.[65] (Under IAEA standards, spent nuclear fuel is considered "self-protecting" if the dose rate at 1 meter is 1 Sv/hour, because anyone trying to steal it would likely receive a lethal radiation dose.) On these grounds, the U.S. MPC&A program did not add any physical-protection security measures to the spent fuel storage. But this is odd because the first nuclear icebreaker, *Lenin,* started up in 1959, so there is probably a significant amount of spent fuel that is no longer self-protecting. Murmansk Shipping Company claimed that the fresh fuel was enriched to only 35 percent, which if true would mean that all or most of the spent nuclear fuel would be LEU. But assuming that some of the fresh fuel originally was enriched to 90 percent, there could be substantial quantities of non-self-protecting HEU in the spent fuel on the storage ships.

## Summary of Findings

If Russia proceeded with its original plan to retire the current HEU-powered icebreakers and build a new generation propelled by the LEU-fueled RITM-200, the fleet could be converted to LEU fuel as the new vessels came on line.[66] Alternatively, if Russia opted to convert the reactors in its existing fleet to LEU fuel, their cassette core design would facilitate the required change in fuel configuration and volume without any modifications to the reactor vessel, enabling another path to phase out HEU fuel relatively rapidly. A Rosatom official mentioned supposed plans to convert the current icebreaker fleet to LEU fuel, but he was unable to provide any further details about the project.

In reality, however, continuation of the lifetime extensions that have proved successful in the past is likely to perpetuate the fleet's use of HEU fuel for a decade or more. The cassette core design requires the entire core to be replaced, so the introduction of LEU fuel would probably not be implemented at least until a necessary refueling. Indeed, it is conceivable that HEU fuel might continue to be used in the current fleet until those ships are decommissioned.

If construction of the LEU-fueled, floating nuclear power plant is successful – employing a reactor based on one used in existing icebreakers – it would prove that conversion of this current icebreaker propulsion reactor to LEU fuel is feasible without any modification to the reactor. The planned use of LEU fuel in the floating plant also means that converting the icebreakers to LEU could yield savings from economies of scale if a common fuel design were standardized.

## Conclusions

The transition from HEU to LEU fuel in Russia's nuclear-powered icebreaker ships seems to be underway. Yet, in their statements, Russian officials seem unconcerned about nuclear proliferation and nuclear terrorism. Security issues and proliferation are dismissed as a problem of the past. Despite two fuel thefts in the region in the 1990s, the risk of an insider stealing weapon-usable nuclear materials is not considered a serious problem by the Russian scientists that were interviewed.

Russia's conversion effort is instead motivated by the potential monetary gain from exporting its reactor designs to non-nuclear weapons states, given that eschewing HEU fuel would avoid much international opposition. While the Russians have no concrete plans to export nuclear icebreakers, their icebreaker fleet is also being used as a testing ground for small modular power-reactor designs that could be implemented on land or barge-mounted facilities.

Another potential economic incentive for converting the nuclear icebreaker fleet to LEU would be the efficiency of unifying the reactor designs for the nuclear icebreaker fleet, nuclear submarines, and floating plants. Assuming economies of scale for fuel fabrication and reactor production, and maintenance and refurbishment, the cost could be decreased for both domestic naval propulsion and exported modular reactors. In order to calculate the benefits of a universal conversion to a standardized LEU reactor, an in-depth study requiring a number of specialists should be undertaken. Aspects that would need to be examined include the cost of HEU vs. LEU fuel, the cost of physical protection for HEU vs. LEU facilities, and the production time of the reactor assemblies. Russian scientists responded positively to the prospect of such proposed research.[67] However, since the number of specialists needed would be quite large, the required funding and organization would be substantial, probably necessitating outside assistance. Foreign governments, international organizations, and NGOs concerned with nuclear security should seriously consider funding such research to facilitate and expedite Russia's phase-out of HEU for civilian naval propulsion, as part of larger global phase-out of HEU for non-weapons purposes, to minimize risks of nuclear terrorism and nuclear proliferation.

## Notes

1 The passage is also known as the North East Passage or Northern Sea Route.

2 Oleg Bukharin, "Security of Fissile Materials in Russia," *Annual Review of Energy and the Environment* 21 (1996): 479–480.

3 Jorn Madslien, "Nuclear waste poses Arctic Threat," *BBC News*, 19 October 2006.

4 Michael A. Champ, V. V. Makeyev, James M. Brooks, Ted E. Delaca, Kay M. van der Horst, and Marilyn Vorela Engle, "Assessement of the Impact of Nuclear Wastes in the Russian Artic," *Marine Pollution Bulletin 35 (1998):* 207–211.

5 Ole Reistad, Morten Bremer Mærli, and Nils Bøhmer, "Russian Naval Nuclear Fuel and Reactors: Dangerous Unknowns," *Nonproliferation Review* 12 (2005): 164.

6 The refueling ship is used to store the fresh fuel cores. This ship also allows for the removal of spent nuclear fuel cores from icebreaker ships.

7 Nicole Nelson, Parker O'Shell, Sheilah Hendrickson, V. Sukhoruchkin, S. Antipov, E. Melkhov, N. Ponomarev-Stepnoi, and Admiral N. Yurasov, "Cooperative MPC&A Enhancements at Russian Navy Sites" Presentation at the Institute of Nuclear Materials Management 42nd Annual Meeting, Indian Wells, CA, 15-19 July 2001, 5.

8 Bukharin, "Security of Fissile Materials in Russia," 479–480.

9 F. Duran, S.H. Conrad, G. N. Conrad, D. P. Duggan, and E. B. Held, "Building A System For Insider Security," *Security & Privacy, IEEE* 7 (2009): 30–38.

10 Nils Bøhmer, Aleksandr Nikitin, Igor Kudrik, Thomas Nilsen, Andrey Zolotkov, and Michael H. McGovern, "The Arctic Nuclear Challenge," *Bellona Report Volume 3* (The Bellona Foundation: 2001): 29.

11 Anatoli Diakov, email to author, stated the fuel is a uranium-aluminum alloy with uranium enrichment around 36 percent.

12 $MW_t$ refers to the thermal power output of the reactor. $MW_e$, commonly used to describe nuclear plants that are used for electricity generation, refers to the amount of power produced by the system that can be used as electricity.

13 Anatoli Diakov, Alexander M. Dmitriev, Jungmin Kang, Alexey M. Shuvayev, and Frank N. von Hippel, "Feasibility of Converting Russian Icebreaker Reactors from HEU to LEU Fuel," *Science and Global Security* 14 (2006): 34–42.

14 M. B. Maerli, S. Børresen, K. Gussgard, and S. Høibråtten, "Criticality Calculations on Russian Ship Reactors and Spent Nuclear Fuel," Norwegian Radiation Protection Authority, 1998.

15 Diakov et al., "Feasibility of Converting."

16 See this book's chapter by Margarita Jimenez.

17 Reaching a maximum U-235 burn-up of 93 percent and an average of 84 percent in mini elements. For a full-scale test, the group was able to reach an average U-235 burn-up of 50.3 percent. A. L. Izhutov, V. V. Alexandrov, A. Ye. Novosyolov, V. A. Starkov, A. V. Sheldyakov, V. Yu. Shishin, V. V. Iakovlev, I. V. Dobrikova, A. V. Vautlin, V. B. Suprun, Ye. F. Kartashov, and V. A. Lukichev, "The Status of LEU U-Mo Fuel Investigation in the MIR Reactor, " RERTR 2009, 31st International Meeting on Reduced Enrichment for Research and Test Reactors, Beijing, China, 1–5 November 2009.

18 "Универсальный атомный ледокол нового поколения," 18 October 2010.

19 Galina Stolyarova, "Nuclear Power Vessel Launched," *The St. Petersburg Times*, 6 July 2010.

20 Dispersion fuel generally refers to an inert matrix surrounding the fissile material. The most common assumption for the fuel being used in the floating nuclear power plants is $UO_2$ dispersed in a silium medium. Vladimir Kuznetsov, email to author, claimed the fuel is a $UO_2$ fuel in a Zr matrix.

21 The testing phase at the Baltiisky shipyard in St. Petersburg started in July 2010. Stolyarova, "Nuclear Power Vessel Launched." G.V. Kulakov, A. V. Vatulin, S. A. Ershov, A. E. Novoselov, V. A. Ovchinnikov, and V. Y. Shishin, "Development of Dispersion Type Fuel Element for Floating Nuclear Power Plant," 2010 LWR Fuel Performance Meeting/Top Fuel/WRFPM. September 2010. Vladimir Kuznetsov, email to author, claimed the fuel was 13 percent and 15.7 percent enriched in two radial zones in the core. The different enrichment zones flatten the flux profile across the core.

22 "Design Features to Achieve Defence in Depth in Small and Medium Sized Reactors," *IAEA Nuclear Energy Series,* No. NP-T-2.2 (2009).

23 Vladimir Kuznetsov, "IAEA Activities for Innovative Small and Medium Sized Reactors (SMRs)," *Progress in Nuclear Energy* 47 (2005): 69.

24 On-site refueling means the reactor can be refueled at its current site. Therefore, the floating nuclear power plants will probably have storage for fresh fuel and pool storage available for spent fuel. The amount of fuel stored in the facility at any time will be several times more than is necessary to fuel the reactor.

25 Effective core life of 70,000 hours.

26 "Status of Small Reactor Designs without On-Site Refueling," *IAEA-TECDOC-1536*. January 2007. Due to the increased enrichment, the reactor plant has a refueling schedule on the same timeline as the maintenance schedule. Therefore, the refueling is scheduled to happen only when the power plant is back in the Russian port getting general maintenance.

27 Translated by Igor Kudrik.
28 The interview was translated and transcribed by Anna Kireeva, the public relationship specialist for Bellona-Murmansk.
29 In an interview with V. Kuznetsov, he claimed the Russian nuclear fleet would need five icebreakers to carry out the necessary functions. Y. Sidorov claimed there will need to be seven nuclear icebreakers dedicated to the Northern Sea Route.
30 Interview with V. Kuznetsov
31 Interview with A. Dmitriev. Interview with V. Kuzentsov.
32 This is stated in a pamphlet about the reactor design and verified via email by V. Kuznetsov.
33 Interview with V. Kuznetsov.
34 Email correspondence with V. Kuznetsov, 9 March 2011.
35 Email correspondence with I. Kudrik, 24 March 2011.
36 Interview with A. Zolotkov, translated by Anna Kireeva.
37 Bøhmer et al., "The Arctic Nuclear Challenge."
38 "Russia as a world leader in nuclear-powered icebreaker fleet construction," *Voice of Russia*, 9 September 2012, http://english.ruvr.ru/2012_09_09/Russia-as-a-world-leader-in-nuclear-powered-icebreaker-fleet-construction/ (accessed on 27 September 2012).
39 Interview with A. Zolotkov.
40 The proliferation-resistance claim is due to the fact that the Mayak reprocessing facility is the only site with the technology to reprocess the spent fuel on a commercial scale. From an interview with A. Nikitin, translated by Igor Kudrik. The amount of plutonium in the spent nuclear fuel is around 1 percent. A large amount of SNF would need to be reprocessed in order to extract enough Pu to make a single nuclear weapon. Therefore, the Russians argue there is a good chance no one will try to extract Pu from this type of SNF for the construction of a nuclear weapon.
41 Interview with A. Nikitin, translated by Igor Kudrik.
42 Email correspondence with Y. Sidorov, 13 February 2011, including information based on consultation with reactor designers from OKBM.
43 Anna Kireeva and Charles Digges, "Russia's Atomflot inks deal to build world's largest nuclear icebreaker," *Bellona*, 24 August 2012, http://www.bellona.org/articles/articles_2012/biggest_icebreaker (accessed on 27 September 2012).
44 Interview with A. Zolotkov.
45 Interview with V. Kuznetsov.
46 Interview with N. Arkhangelskiy.
47 Interview with Y. Sidorov.
48 Email correspondence with V. Kuznetsov, 17 October 2010.
49 Email correspondence with A. Zolotkov, 3 November 2010.
50 Robert Berls, email correspondence, 19 October 2010, claimed the Russians never responded after the proposal was drawn up, so the project was dropped.
51 Interview with V. Kuznetsov.
52 Cristina Hansell Chuen and Ole Reistad, "Sea Fission: Russia's floating nuclear power plants," *Jane's Intelligence Review,* December 2007, 52.
53 A.A. Sarkisov, V. L. Vysotskii, V. P. Bilashenko, V. N. Barinov, A. E. Kiselev, V. G. Markarov, I. N. Kulakov, A. N. Lepekhin, V. V. Petrunin, A. V. Pichugin, and S. V. Krysov, "Expected Radiological and Radioecological Consequences of Operating Floating Nuclear Heat and Power Plants," *Atomic Energy* 104 (2008): 237.
54 V.I. Makarov, B. G. Pologikh, N. S. Khlopkin, F. M. Mitenkov, Yu. K. Panov, V. I. Polunichev, and O. A. Yakovlev, "Experience in Building and Operating Reactor Systems for Civilian Ships," *Atomic Energy* 89 (2000): 697. Sarkisov et al., "Expected Radiological and Radioecological Consequences," 248.
55 Translated by A. Kireeva.
56 Interview with V. Kuznetsov.

57 Interview transcript from A. Zolotkov, translated by A. Kireeva.
58 Interview with N. Arkhangelskiy.
59 Interview with V. Kuznetsov.
60 Email correspondence from Y. Sidorov, 10 March 2011.
61 Interview with O. Reistad.
62 Interview with V. Kuznetsov.
63 Interview with A. Zolotkov.
64 Bøhmer et al., "The Arctic Nuclear Challenge," 36.
65 Nelson et al. "Cooperative MPC&A Enhancements," 3.
66 The construction of the first three vessels powered by the RITM-200 reactor is expected to be completed by 2027. See, Kireeva and Digges, "Russia's Atomflot inks deal."
67 Email correspondence with Y. Sidorov, 10 March 2011. Sidorov offered to organize a group of experts to undertake this project if funding were available.

# 6 Belgium and the Netherlands

## Medical Isotope Production

*Alexander Fay*

A surprise announcement at the 2012 Nuclear Security Summit – in Seoul, South Korea – overturned decades of previous claims and expectations about the feasibility and likelihood of phasing out Europe's use of weapons-grade, highly enriched uranium (HEU) for the production of medical isotopes.[1] Two of the three largest producers of medical isotopes in the world are based in Europe: Covidien, in the Netherlands; and the National Institute for Radioelements (IRE), in Belgium. These two companies traditionally produce roughly half of the world's supply of Molybdenum-99 (Mo-99), the isotope used to make most of the world's radiopharmaceuticals that are used in tens of millions of diagnostic procedures annually. Nuclear security advocates for years had urged the two companies to convert their production processes from HEU to low-enriched uranium (LEU), which is unsuitable for nuclear weapons. But until the summit, the companies had argued publicly that although such conversion was technically feasible, it would require at least six years and might not be economically feasible. Then, on 26 March 2012, both Belgium and the Netherlands signed an agreement to phase out HEU use entirely within three years. The pact was also signed by the United States, which supplies enriched uranium, and France, home to the firm that fabricates the enriched uranium into targets that the two companies then irradiate to produce medical isotopes. The agreement declared: "Belgium, the Netherlands, and France, in cooperation with the United States, reaffirm their determination to support conversion of European production industries to non-HEU-based processes by 2015."[2]

This chapter explores the background and implications of this landmark agreement. It finds that the timing of the decision was compelled by an impending shortage of HEU in Europe, and a U.S. policy of encouraging conversion by restricting exports of HEU. But the Europeans' decision that conversion would be feasible, technically and economically, was also based on years of research and development by laboratories and isotope producers around the world, and the specifics of the two companies' facilities, as detailed below.

## Background

In 2008, in an untimely coincidence, two nuclear reactors that had been producing most of the global supply of Mo-99 suffered simultaneous extended shutdowns.

The outages of the NRU in Canada and the HFR in the Netherlands, and the resulting global shortage of medical isotopes, focused world attention on the fragile infrastructure to produce this vital resource. Since then, various companies, countries, inter-governmental organizations, and advocacy groups have undertaken to secure the supply and stabilize the economy for Mo-99. New and historically small-scale producers have set their sights on increasing production to capture more of a market that historically has been dominated by a small number of large producers. Attention has also focused on the use of HEU in the Mo-99 production process. In light of the perceived proliferation and terrorism risks of civilian HEU use, many expert organizations have advocated phasing out the use of this material in the isotope production process. These organizations include the U.S. National Nuclear Security Administration (NNSA), the International Atomic Energy Agency (IAEA), the Nuclear Energy Agency of the Organization for Economic Cooperation and Development (OECD NEA), the U.S. National Academy of Sciences (NAS), and the non-governmental Nuclear Threat Initiative (NTI). They have dedicated technical, economic, and academic resources to assess the role of HEU in the Mo-99 supply chain, and are working towards converting isotope production processes to use low enriched uranium (LEU) or to avoid enriched uranium altogether.

For Covidien and IRE, the majority of practical work towards conversion has resulted from prodding and assistance by the NNSA's Global Threat Reduction Initiative (GTRI). Belgium and the Netherlands first endorsed the principle of conversion at the first ever Nuclear Security Summit (NSS) – in 2010, in Washington, DC – but neither government followed up by providing tangible support to their domestic companies to facilitate such conversion. In 2011, however, as the shortage of HEU in Europe became apparent, Dutch government ministries started increasing their attention to the issue.

Traditionally, the United States and Europe have accounted for the overwhelming majority of global demand for Mo-99, which has been met by a small number of suppliers. As of 2009, three producers alone – Covidien, IRE, and Canada's Nordion – supplied nearly 90 percent of the global demand of 12,000 weekly 6-day Ci,[3] although the relative contribution by each supplier often changed weekly. Smaller, regional suppliers, including Argentina's Comisión Nacional de Energía Atómica (CNEA), South Africa's NTP, and the Australian Nuclear Science and Technology Organisation (ANSTO), were not involved in the global supply network until very recently.

Covidien, IRE, and Nordion all produce Mo-99 by the same general process: irradiation of HEU targets, thereby creating some Mo-99 when uranium fissions; dissolution of the irradiated targets into a liquid solution; and then extraction of Mo-99 from the liquid. The purified Mo-99 is then provided to manufacturers of "generators," which are used to extract the isotope produced when Mo-99 decays, Tc-99m, which is the most commonly used medical radiopharmaceutical. The Tc-99m generators are then distributed to medical facilities. When the isotope is needed, a physician extracts Tc-99m using the generator, then prepares the kit appropriate for the diagnostic exam being administered to the patient.

The importance of Tc-99m to modern medicine, and the isotope's short half-life of only six hours, makes a reliable supply of Mo-99 essential. But maintaining this supply is difficult because Mo-99 cannot be stockpiled, given that it also has a relatively short half-life of only 66 hours. Accordingly, Mo-99 must be produced steadily and without major interruption – a major challenge confronting the nonproliferation community's efforts to phase out HEU from isotope production. Another challenge is the proprietary nature of some isotope production processes, which makes it difficult to assess claims about alleged obstacles to conversion. In addition, unlike global efforts to phase out HEU from other types of nuclear facilities, such as research reactors and critical assemblies, advocates cannot call for consolidation or decommissioning of medical isotope production facilities because that would risk public health. Accordingly, the strategy of the nonproliferation community towards medical-isotope production has been to promote the feasibility of satisfying the Mo-99 supply without the use of HEU, by converting existing facilities to use LEU, or by building new facilities that do not require HEU.

Nonproliferation advocates have tried for more than two decades to persuade large-scale isotope producers to work towards converting their production processes from using HEU to LEU targets. Past conversion by Argentina's CNEA and South Africa's NTP, and the resistance to conversion by Canada's Nordion, are discussed in separate chapters of this book. This chapter assesses the feasibility of conversion by the Netherlands's Covidien and Belgium's IRE, and determines the state of conversion efforts by both companies.

The next section of this chapter reviews the literature on Mo-99 production and conversion. Following that is an explanation of the study's methodology, including interviews with European officials. The research findings are then detailed and summarized. The chapter concludes with recommendations to promote the phase out of HEU from production of medical isotopes in Europe and beyond.

## Literature Review

The literature regarding Mo-99 production is substantial. In addition to large-scale assessments of the global Mo-99 market by the U.S. National Academies and the OECD High-level Working Group on Security of Supply of Medical Radioisotopes (HLG-MR), there exists an array of articles regarding the state of LEU target research, proceedings from the various IAEA working groups, and publications by the producers. As a result, the production side of the Mo-99 market is quite well documented and understood. The major exception is accurately assessing the feasibility of conversion, due to the proprietary nature of Mo-99 production processes. It should also be noted that the Mo-99 production industry is highly dynamic and publications can be rendered out-of-date rather quickly. The technical capabilities of the producers and their public position on conversion, the number of producing countries and companies, the state of U.S. policy, and the state of LEU target research have all changed during the last few years.

As early as the 1980s, investigations showed that LEU targets could replace HEU targets for the production of Mo-99 with minimal penalty to purity or to the yield of Mo-99 from the U-235 in the target.[4] Some concerns with production from LEU targets were identified as well, but these issues – such as increased waste volume and increased production of Pu-239 – have not proved prohibitive.[5] Higher density uranium targets, made from uranium silicides ($U_3Si_2$) and uranium foils, have been produced and tested.[6] LEU dispersion targets have been proven viable and are already in use commercially, where they are processed using the same chemical techniques as HEU targets.[7] LEU targets are used by commercial Mo-99 producers in Argentina, Australia, and South Africa.[8] Alternative techniques to produce Mo-99 without HEU have also been investigated, including "activation" of Mo-98 (i.e., absorbing a neutron to become Mo-99) in a traditional nuclear research or power reactor, production in future liquid LEU-core reactors, or use of a particle accelerator that would avoid the need for a reactor entirely. But none of these alternatives is yet able to produce economically a major share of global demand.[9]

Estimates of the costs associated with conversion to LEU targets suggest that conversion is viable with a minor increase in the price per dose of the delivered radiopharmaceutical. One cost estimate for the conversion of Covidien's facilities in the Netherlands is $10 million, which could be offset by an increase in the price of the radiopharmaceutical of less than 1 percent.[10] It should be noted, however, that industry representatives claim that a prominent 2009 study by the U.S. National Academy of Sciences underestimated these conversion costs, as well as the regulatory hurdles incurred by conversion.[11] Subsequent estimates of the costs along the supply chain were made in 2010 by the OECD HLG-MR in its interim report.[12] This report confirmed the NAS assessment that even substantial increases in costs to the producers would not substantially increase the cost to patients of the delivered radiopharmaceutical, although the OECD report did not explicitly estimate the costs of conversion.[13]

Covidien and IRE currently irradiate targets at a number of reactors, but primarily at the HFR in the Netherlands and BR-2 in Belgium. Each company processes the targets at its respective facilities in Petten, Netherlands and Fleurus, Belgium. Both companies use annular or plate targets, with U-Al alloy constituting the target meat,[14] and process the targets by alkaline dissolution.

NRG, the operator of HFR, has announced plans to construct the follow-on Pallas reactor at a site that would not be licensed for the presence of HEU, thereby requiring Covidien to convert to LEU targets to produce Mo-99 there. In light of the originally projected 2016 start-up date for the Pallas reactor, Covidien had said it would first develop LEU production techniques for targets to be irradiated in Pallas and only afterwards investigate applying those techniques to convert to LEU its existing targets that are irradiated at HFR and other reactors.[15] But the delay in the projected start-up of Pallas until 2018–2020 has focused attention on converting the existing targets to LEU. Covidien has not provided specifics on its LEU conversion research. IRE had previously stated publicly that it was not pursuing conversion, but the National Academy of Sciences concluded that conversion by Covidien would likely compel IRE to convert as well.[16]

The first U.S. law to address the potential phase-out of HEU in the production of Mo-99 was the Schumer Amendment to the 1992 Energy Policy Act, which required that foreign producers commit to convert to LEU targets as soon as possible in order to receive HEU exports from the United States in the interim prior to conversion.[17] These restrictions were lifted 13 years later by the Burr Amendment to the 2005 Energy Policy Act, which exempted Canada and four European countries from the provisions of the Schumer Amendment regarding HEU exports for isotope production.[18] More recently, the American Medical Isotope Production Act would prohibit HEU exports within 7–13 years, depending on the availability of Mo-99 produced without HEU, while also authorizing funds for development of a domestic industry to produce Mo-99 without HEU.[19] The bill was reported by the U.S. Senate in 2011, after a leading opponent retired,[20] and was enacted in January 2013.

Conversion of isotope production was also a focus of the first ever Nuclear Security Summit, in 2010, in Washington, DC. A provision of the work plan that emerged from the event declares:

> Participating States, as appropriate, will collaborate to research and develop new technologies that require neither highly enriched uranium fuels for reactor operation nor highly enriched uranium targets for producing medical or other isotopes, and will encourage the use of low enriched uranium and other proliferation-resistant technologies and fuels in various commercial applications such as isotope production.

Two years later, the final communique of the following summit stated similarly, if more succinctly: "We also encourage States to promote the use of LEU fuels and targets in commercial applications such as isotope production." Both documents were agreed upon by all states participating in the NSS, including Belgium and the Netherlands.[21]

In light of the substantial nature of the relevant literature, many important questions regarding conversion have at least been asked, if not already partially answered. But these questions warrant revisiting in order to provide the most up-to-date information. The remainder of this chapter provides updated information on a number of the existing questions and provides new insight by addressing questions that have not been answered elsewhere in the literature.

## Methodology

Field research was conducted in order to answer the following questions relevant to isotope production in Belgium and the Netherlands:

- What is the current state of research on, and any plans for, conversion by IRE and Covidien? If either company is pursuing conversion, which type of target would be utilized? Would processing and irradiation occur at the same facilities as current production?
- From the perspective of the producers, are there prohibitive technical or economic obstacles?
- How many processing lines do Covidien and IRE utilize? If either company uses only one, would they construct additional lines during conversion, to avoid interruption in production? If either company uses two lines, would conversion occur incrementally, avoiding interruption in production? Would either company ultimately maintain both HEU and LEU processing lines?
- Who are the HEU suppliers for Covidien and IRE? Is the reliability or availability of future supply a concern?
- How much HEU per year do Covidien and IRE consume? How much HEU does each company have stockpiled for target production? What is done with the spent target material?
- If Covidien begins using LEU targets to produce Mo-99 at the Pallas reactor when that facility opens, would it also convert to LEU its existing targets for irradiation at HFR and other reactors, or would it simultaneously irradiate and process both HEU and LEU targets? Does it plan to build a new process line for LEU targets or to convert an existing line?
- Are there regulatory hurdles in either country that are a deterrent to conversion, even if that proved economically viable? If so, what reform could be enacted to overcome such hurdles?
- What efforts has the United States made to encourage conversion by these two European producers?
- Is Covidien or IRE concerned about a potential U.S. tariff on, or prohibition of, the import of isotopes produced with HEU? Does either possibility affect their plans to convert?

In addition to the producers themselves, related entities were potentially useful in answering these questions. Foreign government officials, research organizations, regulatory agencies, reactor operators, Tc-99m generator manufacturers, and others are responsible for aspects of the Mo-99 production infrastructure. Inquiries were directed to several such relevant organizations in Belgium and Netherlands, as well as to the NNSA in the United States.

In-person interviews were conducted in Belgium with Dominique Moyaux of IRE, and in the Netherlands with Dr. Harrie Seeverens of the Ministry of Health, Welfare, and Sport. Discussions were held with Dr. Parrish Staples and Rilla Hamilton of the NNSA when they visited our research project at the University of Texas, and a follow-up phone interview was conducted with Ms. Hamilton. Email exchanges were conducted with Dr. A.A. Sameh, the scientist who originally developed Covidien's Mo-99 recovery process, Fred Wijtsma of NRG, Luis Barbosa of Covidien, and Frank von Hippel of Princeton University.

The proprietary nature of the producers' operations and business strategy posed a partial obstacle to this research. Representatives at both companies were either unable or unwilling to answer certain questions. NNSA officials were similarly unable to answer some questions in consideration of their collaborative efforts with the Mo-99 producers, and other organizations deferred some questions to the producers. Covidien representatives were unwilling to answer specific research questions, but IRE officials were more forthcoming. Accordingly, many of this chapter's technical findings are based on information obtained from IRE.

## Findings

### *Conversion Efforts*

IRE is collaborating with the NNSA on conversion research and has also taken part in the IAEA Conversion Planning working group. As part of the working group, IRE and Covidien have agreed to finalize the bounding criteria for LEU foil targets, recommend research to further develop LEU $U_3Si_2$ targets, and provide information on the waste management regulations they face. Based on the discussion of IRE's processing capabilities below, it does not seem that new facilities would be required for conversion. IRE estimated that it could require 6 to 7 years to establish a complete LEU-based Mo-99 production infrastructure if conversion were undertaken with uranium foil or silicide targets. The estimated timeframe would be shorter if conversion utilized $UAl_2$ dispersion targets.[22]

### *Technical and Economic Obstacles*

For IRE in Belgium, each of three possible LEU target materials – aluminide, silicide or foil – has technical and/or economic drawbacks, but the company appears to be gravitating towards LEU foil targets. Regardless of which material is chosen, the company will dissolve the targets in an alkaline solution because the alternative of an acidic dissolution would increase gas emissions, which the Belgian government insists should be reduced, according to Dr. Moyaux. The option of $UAl_2$ LEU dispersion targets – as utilized by South Africa's NTP and Argentina's CNEA – would be easiest technically on the front and back end. The dispersion matrix is readily available for fresh targets and the processing chemistry for irradiated targets would require only minimal changes to IRE's existing dissolution process. But a larger number of targets would have to be irradiated to maintain output of Mo-99 because the density of $UAl_2$ in LEU targets is not yet high enough to yield the same amount of Mo-99 per target as current HEU targets. This increase in the number of targets would in turn necessitate finding more irradiation positions in reactors and more or different transportation containers, which currently hold only three irradiated targets each.

Accordingly, IRE is also exploring the LEU alternatives of silicide or uranium foil. Either option would entail significant changes to the non-uranium composition of the target, necessitating substantial modification of both the processing chemistry for irradiated targets and the management of waste that would have different composition and possibly volume. Moreover, IRE is concerned that applying its current processing techniques, originally developed by Dr. Sameh, to $U_3Si_2$ LEU targets could damage its hot cells, so additional modification to equipment might be necessary.

In light of these concerns with aluminide silicide, IRE officials appear to prefer the third conversion option of an annular target with an LEU foil pressed between cylindrical aluminum cladding. But Dr. Moyaux noted that this option also poses substantial challenges beyond modifying the dissolution and purification processes for irradiated targets. Indeed, the foil targets have not yet even been qualified for large-scale reactor irradiation. Moreover, IRE has not yet confirmed a supplier of the raw material for LEU foils nor a large-scale fabricator to transform this material into targets.

The NNSA reports that foil targets have already been successfully developed and that work has begun at Argonne National Laboratory on the chemical dissolution of the targets and qualifying them for commercial irradiation.[23] According to Argonne and IRE, the Mo-99 yield of the LEU foil targets is equivalent to that of current HEU targets.[24] The LEU foil targets would initially cost more for the isotope producers but this cost difference may decline over time, and in any case would not greatly affect the price of the delivered radiopharmaceutical to patients.[25] Licensing of the targets by national regulators, after they are qualified and accepted by the producers, may take some months but is not seen as a prohibitive step in the conversion process.

The economic impact of conversion is difficult to estimate precisely, in part because the global supply of Mo-99 has been unstable over the past few years, leading to large price fluctuations. Moreover, governments in countries that are home to major isotope producers have only recently become aware of the extent to which subsidies affect the Mo-99 market.[26] By paying for operations and maintenance of the research reactors at which targets are being irradiated, the governments of Belgium and Netherlands, among other countries, are effectively subsidizing Mo-99 production using HEU. (Seeverens estimates that the actual value of the reactors to the isotope producers may be 5 to 7 times higher than they pay to irradiate their targets.) Such subsidies are the main reason that the global Mo-99 market is currently deemed unhealthy and unsustainable, inhibiting the emergence of non-subsidized producers that avoid HEU.[27] After publication of the HLG-MR report in 2011, the group continues to advocate scaling back, subsidies, which should facilitate conversion by existing producers to LEU targets and the emergence of new producers that avoid HEU. Cooperation between producers may be necessary to evaluate and properly manage the post-subsidy market, to the extent this is permitted by anti-trust restrictions.

### *Conversion Would Not Interrupt IRE Production*

According to Moyaux, conversion to LEU targets at IRE would not cause an interruption in production. IRE maintains two processing lines, one consisting of a series of three hot cells and the other a series of six cells. Different steps of the dissolution and purification process are conducted in each cell. In 2010, production took place exclusively on the first line until refurbishment of the second line was completed in September of that year. Conversion to an LEU target would require modification of processes in the first three cells of one line, but this could be done without interrupting production by shifting that work to the second line that was not currently in use. The processing cycle for HEU targets would then utilize cells 1–3 of line 2, and cells 4–6 of line 1. In the meantime, decontamination and modification of line 1's first three cells would take approximately 18 months, at which point all processing of LEU targets could take place on line 1. (If desired, the second line could then be converted.)

A similar conversion sequence would *not* be possible at Covidien, which utilizes both of its processing lines simultaneously. Thus, conversion at Covidien would likely require construction of a new processing line, or a temporary reduction in operations, although this was not confirmed by Covidien.

### *HEU Supply*

For at least two decades, IRE's HEU has come from a European stockpile originally supplied by the United States. In interviews in early 2011, IRE officials did not comment on concerns about the reliability of future HEU supply, nor would they provide numbers regarding their HEU stockpile or consumption rate. According to Moyaux, spent target material is reprocessed after 2–3 years for the purposes of strontium extraction, but not to recover HEU for reuse in targets. IRE would not provide further details on waste disposal.

### *The Role of Pallas in Conversion Planning*

Even prior to the 2012 agreement in Seoul, it was widely agreed that the start-up of operations and isotope production at the forthcoming Pallas reactor would mark a deadline for Coviden's conversion to LEU targets,[28] because the new reactor site would not be licensed for HEU. It is not known if IRE also intends to irradiate targets in the Pallas reactor. The projected start of operations at Pallas has been delayed until at least 2018–2020, due to funding shortages.[29]

### *Government Involvement*

The NNSA has worked with international producers to research LEU-based technology to produce Mo-99 and has provided financial support for such conversion. The White House retroactively acknowledged this effort on 7 June 2012, declaring that it would be "continuing to provide support to international producers to assist in the conversion of Mo-99 production facilities from HEU to LEU."[30] But the NNSA has declined to comment on specific collaborations with foreign companies.

When asked about support from the United States, Moyaux stated that IRE is receiving all the assistance it requires in terms of conversion research, but in reality the extent to which the NNSA can work with IRE is limited. GTRI does not permit techniques it funds to be patented, so any modification or overhaul of the purification process to extract Mo-99 from dissolved targets would have to be developed solely by IRE.[31] Foreign Mo-99 producers have requested that the NNSA facilitate their conversion by proving the viability of all LEU-based processing steps with the exception of purification.[32]

According to Seeverens, the Netherlands government, despite its rhetorical support for HEU minimization, has taken no practical steps to facilitate or incentivize Covidien's conversion to LEU targets. Dutch government ministries have recently increased their attention to conversion, but it remains to be seen if this will translate into action. As of 2011, there was still disagreement within the Dutch Ministry of Foreign Affairs about how to implement the pledges made by the country to HEU minimization at the first Nuclear Security Summit in 2010. Likewise, as of 2011, Belgium's government did not appear to have taken any concrete steps to encourage conversion.

### Quadripartite Agreement (by Alan J. Kuperman)

When Belgium and the Netherlands committed – in March 2012, at the 2nd Nuclear Security Summit, in Seoul, South Korea – to convert their isotope production from HEU to LEU targets by 2015, it caught many observers by surprise, especially in light of IRE's and Covidien's past protestations that conversion would take much longer and might be economically unfeasible. The agreement culminated more than a year of sensitive negotiations between the two countries and the United States and France. The short explanation of why Belgium and the Netherlands committed to convert quickly to LEU is that Europe had run out of HEU, and the United States insisted on such a commitment as a condition of temporarily supplying HEU to avert an interruption in the production of medical isotopes. The full explanation, however, is far more intriguing.

For the preceding two decades, the two European companies had produced Mo-99 using targets made of stocks of HEU within Europe that likely originated from the United States, Russia, and/or United Kingdom.[33] Under U.S. law from 1992 to 2005, the companies were not eligible for fresh exports of HEU because they had not committed to convert to LEU targets as soon as possible. But under the U.S.-Euratom nuclear agreement, the companies could have utilized any available stocks of U.S.-origin HEU that had been previously exported to Europe for other peaceful uses. Much of Europe's civil HEU was stored in France at the nuclear company CERCA,[34] which fabricates targets and fuel for research reactors. In particular, CERCA stored hundreds of kilograms of HEU that Russia had exported to Europe in the late 1990s for use as fuel in three research reactors, including the RHF-Grenoble in France.[35]

Over time, CERCA exhausted its stock of surplus HEU to fabricate targets for the two isotope-production companies. So, the French fabricator resorted to a measure that may have violated its contracts. CERCA decided to fabricate targets from Russian-origin HEU that was not surplus but had been exported to Europe explicitly to fuel the RHF-Grenoble research reactor that still required HEU fuel. As a result, the reactor operator needed to acquire more HEU for future operations than it had anticipated. It already had applied for an HEU export from the United States but now would need to increase that request.

The amount of HEU misused by CERCA is not insubstantial – an estimated 70 kg – nearly three "significant quantities" by IAEA standards. This estimate is based on the difference between the amounts of HEU requested by RHF-Grenoble before and after the misuse came to light. Beforehand, in December 2010, the reactor operator requested an export license from the U.S. NRC for 116 kg of HEU for fuel.[36] Afterward, in October 2011, it amended the license application to increase the amount of HEU to 186 kg.[37] There is no obvious explanation for this increase other than CERCA's misuse of Russian HEU previously exported to Europe for use as fuel in the RHF-Grenoble.

The incident gave U.S. nuclear-security officials additional leverage in their negotiations, which they used to promote HEU minimization. Europe now required additional HEU from the United States not only to continue producing medical isotopes but also to continue operating one of its highest-performance reactors.

The United States requested three concessions from Europe. First, European officials should provide additional information to reconcile inventories of previously exported U.S.-origin HEU in Europe, to help ensure security and protect against misuse. Second, CERCA should return to the United States most of its remaining fresh "scrap" HEU (that is, remnants from previous fabrication of fuel and targets), which was a security risk and conceivably could have been reconstituted into a stock of fresh HEU to evade U.S. export controls. (The amount of HEU in scrap form to be returned to the United States reportedly exceeds the estimated 70 kg of Russian HEU that CERCA misused.) Third, the two European isotope producers would have to convert from HEU to LEU targets expeditiously.

European officials reportedly agreed to all three U.S. demands (although some details still may be subject to negotiation), enabling the quadripartite agreement to be announced in Seoul in March 2012. The same month, the U.S. NRC approved the license amendment to increase the amount of HEU for export to France to fuel the RHF-Grenoble, substituting for the previously misused Russian HEU. Two months later, in May 2012, IRE applied to the NRC for the export of 6 kg of HEU for targets to produce medical isotopes, reportedly enough for about one year's production.[38] Similarly, in December 2011, Covidien had applied for export of 10 kg of HEU for targets, reportedly about one year's worth for that company, and this license was approved in July 2012.[39] Despite these short-term requests for HEU, there is evidence that the European producers are making efforts to convert isotope production from HEU to LEU targets, in accordance with the quadripartite agreement. In May 2012, for example, a Dutch research laboratory applied to the U.S. NRC for an export of LEU, indicating that its use could include fabrication of "targets to be irradiated in European research reactors to produce Mo-99."[40]

## Summary of Findings

Field work has answered some but not all of the research questions originally posed. Though the companies are not specific about the state of their conversion research, it appears that both Covidien and IRE are preparing for eventual conversion to LEU targets. For IRE, conversion should be possible without an interruption in production. Based on interviews at IRE and the state of research at Argonne, LEU foil technology seems to be the preferred replacement for its HEU targets. For Covidien, the start of operations at the forthcoming Pallas reactor seems to represent a conversion deadline. However, the projected date of the facility's completion has been pushed back to 2018–2020, so both Covidien and IRE aim to convert to LEU targets prior to the opening of Pallas.

At the 2012 Nuclear Security Summit, the countries that are home to the two isotope producers and their fabricator of targets – Belgium, the Netherlands, and France – agreed to eliminate HEU from Mo-99 production by 2015. They did so because of a shortage of HEU in Europe and U.S. insistence on such an agreement as a condition of supplying HEU for targets prior to conversion. The economic impact of conversion is difficult to estimate precisely, in part because the global supply of Mo-99 has been unstable over the past few years, leading to large price fluctuations. This makes it difficult to verify claims by some producers that conversion is not yet economically viable.

The NNSA has ongoing projects with both of the European producers to support their conversion research. In addition, various working groups have been established to sustain collaboration between producers and outside organizations. But the timing of conversion has been driven mainly by U.S. policy, rather than European domestic policies to eliminate HEU from the production process. The construction of Pallas probably would have eventually compelled at least Covidien to convert to LEU targets, but probably not until 2020, whereas U.S. policy on HEU supply is expected to compel conversion by 2015.

## Conclusions

Statements by Covidien,[41] and the ongoing efforts of IRE, indicate that two of the world's three biggest isotope producers recognize the inevitable shift away from using HEU to produce Mo-99. Both Covidien and IRE are currently working towards conversion, based on their collaborative efforts with the NNSA and participation in the OECD and IAEA working groups, though the extent to which research and development is being conducted on their proprietary processes is still unknown.

It is also clear that the construction of facilities at Pallas is a deadline for conversion, at least by Covidien. That company is likely to convert in order to begin irradiating targets at Pallas. The probable shutdown of HFR after Pallas goes online would likely force IRE to convert as well, if it has not already at that point, in order to continue using irradiation facilities in the Netherlands. When Pallas was originally estimated to open in 2016, it appeared unlikely that both companies would convert completely prior to the start of production at that facility. However, the recently announced delay in opening Pallas to 2018–2020 makes it likely that both companies will convert prior to that time, especially in light of their quadripartite agreement at the 2012 Nuclear Security Summit.

In the wake of the 2013 enactment of the American Medical Isotope Production Act, the start of the 7–13 year window for ceasing HEU exports should coincide closely with the opening of Pallas, providing further incentive for both European producers to convert by then. Although the agreements at the first two Nuclear Security Summits are encouraging signs that the governments of Belgium and Netherlands are committed to support conversion, neither has taken concrete action similar to that of the United States, such as developing a policy restricting HEU use. The remaining questions, then, are whether conversion will precede the opening of the Pallas reactor, and by which route these two major producers will choose to convert.

The global economy for Mo-99 is in a state of flux. Prospective producers and existing small-scale producers are both working to become large-scale suppliers. In the wake of temporary shutdowns and impending permanent shutdowns of some reactors, additional reactors like the MARIA in Poland are being utilized for target irradiation. Countries are coordinating through multilateral organizations to stabilize the reliability of supply and make the Mo-99 market sustainable. In the absence of a mandate or incentives to convert, the priority of producers would likely remain fixed on maintaining a profitable Mo-99 supply chain rather than accelerating their conversion efforts. Covidien and IRE, and their host governments Belgium and the Netherlands, now have incentive to convert to LEU targets because they have exhausted their HEU supplies and the United States has utilized its leverage as the main global HEU supplier to demand conversion. But other producers in countries with HEU stocks, notably Russia, may continue to produce, and indeed expand production, using HEU – in the absence of additional incentives for conversion.

## Recommendations

It is important for the future of conversion efforts to implement faithfully the recently enacted American Medical Isotope Production Act that includes a phase-out of HEU exports for medical isotope production. With Nordion recently entering an agreement to distribute worldwide the Mo-99 produced in Russia, it would also be beneficial to work with Russia to phase out its use of HEU for Mo-99 production. Otherwise, Russia's expanding production of Mo-99 could undermine HEU minimization efforts.

It could also be fruitful to apply some pressure to the participants of the Nuclear Security Summit to start crafting policies to motivate conversion by producers, such as a tax or tariff on Mo-99 produced with HEU. According to Dr. Seeverens, conversion will happen quickly once proper policy is in place. Waiting until the end of the decade for the completion of Pallas before converting would represent an unnecessary delay.

Research on the technical feasibility of conversion and the development of LEU-based targets and processing techniques has been ongoing for decades and will assuredly continue to be instrumental in the shift away from HEU-based Mo-99 production. Due to the recent supply shortage, an increasing amount of research is being conducted on the Mo-99 economy as well, but only a small portion of this research is focused on conversion. More work should be conducted to assess the economic impact of conversion, especially considering other changes likely to occur in the supply chain in the near future, and to address producer concerns about cost increases resulting from conversion. If such research is undertaken, it would be crucial to include the major producers so as to minimize the speculative nature of cost estimates and economic predictions.

Collaboration between interested organizations – such as the OECD, IAEA, and U.S. government – has also been an important part of the conversion effort. Collaboration between these groups should continue with an increased focus on the economic effects of conversion and ways to address those effects. Follow-up meetings to the August 2010 Conversion Planning event organized by the IAEA could maintain this collaboration, as have the continued efforts of the HLG-MR. Additionally, information from Covidien regarding its technical capabilities and conversion plans, similar to that provided by IRE, would provide a clearer picture of how the producers can best be supported in their efforts to convert.

In the long-term, continued reliance on Tc-99m/Mo-99 may entail development of alternative production technologies. Research on these options, and the diversification of the global Mo-99 supply regime to include these technologies once mature, can enable future production of Mo-99 that not only is reliable but avoids the security risks associated with highly enriched uranium.

## Notes

1 The author would like to thank Dr. Dominique Moyaux and Dr. Harrie Seeverens for their hospitality and patience in facilitating travel to Belgium and the Netherlands. The author would also like to thank Rilla Hamilton for her time answering questions. Finally, the author would like to acknowledge the representatives at other organizations who were kind enough to respond to inquiries.

2 "Belgium-France-Netherlands-United States Joint Statement: Minimization of HEU and the Reliable Supply of Medical Radioisotopes," White House, Office of the Press Secretary, 26 March 2012, http://www.whitehouse.gov/the-press-office/2012/03/26/belgium-france-netherlands-united-states-joint-statement-minimization-he (accessed on 21 June 2012).

3 OECD, "The Supply of Medical Radioisotopes: Interim Report of the OECD/NEA High-level Working Group on Security of Supply of Medical Radioisotopes," 2010.

4 A.A. Sameh and H.J. Ache, "Production Techniques of Fission Mo-99" (paper, Fission Molybdenum for Medical Use, Karlsruhe, Germany, 13–16 October 1987). G. F. Vandegrift, D. J. Chaiko, R. R. Heinrich, E. T. Kucera, K. J. Jensen, D. S. Poa, R. Varma, and D. R. Vissers, "Preliminary Investigations for Technology Assessment of Mo-99 Production From LEU Targets," meeting on Fission Molybdenum for Medical Use, Karlsruhe, Germany, 13–16 October 1987.

5 Vandegrift et al., "Preliminary Investigations," 110. Frank Von Hippel, "Feasibility of Eliminating the Use of Highly Enriched Uranium in the Production of Medical Radioisotopes," *Science and Global Security* 14, 2–3 (2006): 151–162.

6 C. Conner, M. W. Liberatore, A. Mutalib, J. Sedlet, D. Walker, and G. F. Vandegrift, "Progress in Developing Processes for Converting Mo-99 Production from High- to Low-Enriched Uranium," International Meeting on Reduced Enrichment for Research and Test Reactors, Sao Paolo, Brazil, 18 October 1998.

7 Steven van der Marck, "The Options for the Future Production of the Medical Isotope Mo-99," *European Journal of Nuclear Medicine and Molecular Imaging* 37,10 (2010): 1820.

8 Alan J. Kuperman, "The Global Threat Reduction Initiative and Conversion of Isotope Production to LEU Targets," International Meeting on Reduced Enrichment for Research and Test Reactors, Vienna, Austria, 7 November 2004.

9 Dewi M. Lewis, "(99)Mo supply – the times they are a-changing," *European Journal of Nuclear Medicine and Molecular Imaging* 36,9 (2009): 1371–1374.

10 Von Hippel, "Feasibility of Eliminating," 156.

11 Roy W. Brown, "Global $Mo^{99}/Tc^{99}m$ Supply: Current State and Future Opportunities," presentation, Covidien User Meeting, 15 June 2009.

12 OECD, "The Supply of Medical Radioisotopes," 2010.

13 Harrie Seeverens, "The Economics of the Molybdenum-99/Technetium-99m Supply Chain," *Dutch Journal of Nuclear Medicine* 32,4 (2010): 606.

14 National Academy of Sciences, *Medical Isotope Production Without Highly Enriched Uranium* (Washington D.C.: National Academies Press, 2009), TABLE 2.2, 24.

15 National Academy of Sciences, *Medical Isotope Production*, 120.

16 National Academy of Sciences, *Medical Isotope Production*, 122.

17 Alan J. Kuperman, "Bomb-grade Bazaar," *Bulletin of the Atomic Scientists* 62,2 (2006): 44–50.

18 National Academy of Sciences, *Medical Isotope Production*, 12. The four European countries are Belgium, France, Germany, and the Netherlands.

19 United States, House of Representatives Subcommittee on Energy and Environment. "American Medical Isotopes Production Act of 2009, Section by Section Summary," http://democrats.energycommerce.house.gov/Press_111/20091021/HR3276-section-by-section.pdf (accessed on 27 September 2012). "Text of S.99: American Medical

Isotopes Production Act of 2011," Civic Impulse, http://www.govtrack.us/congress/billtext.xpd?bill=s112-99 (accessed on 27 September 2012).

20 "Sen. Bond blocks debate on American Medical Isotope Production Act," *Columbia Missourian*, 22 July 2010, http://www.columbiamissourian.com/stories/2010/07/22/bond-blocks-bill-expand-nuclear-medicine-tests/ (accessed on 27 September 2012).

21 "Nuclear Security Summit at a Glance," Arms Control Association, http://www.armscontrol.org/factsheets/NuclearSecuritySummit (accessed on 27 September 2012).

22 Dominique Moyaux, Production Manager at IRE, interview by author, March 2011.

23 Rilla Hamilton, Policy Analyst at NNSA, interview by author, February 2011.

24 George Vandegrift, "HEU vs. LEU Targets for Mo-99 Production – Facts and Myths," presentation, Symposium on the Minimization of Highly Enriched Uranium in the Civilian Sector, Oslo, Norway, 18 June 2006.

25 Hamilton, interview, February 2011.

26 Harrie Seeverens, Policy Advisor, Netherlands Ministry of Health, Welfare, and Sport, interview by author, March 2011.

27 Hamilton, interview, February 2011.

28 Seeverens, interview by author, March 2011. Fred Wijtsma, Reactor Manager, NRG, email to author, February 2011.

29 Seeverens, interview by author, March 2011.

30 "FACT SHEET: Encouraging Reliable Supplies of Molybdenum-99 Produced without Highly Enriched Uranium," White House, Office of the Press Secretary, 7 June 2012, http://www.whitehouse.gov/the-press-office/2012/06/07/fact-sheet-encouraging-reliable-supplies-molybdenum-99-produced-without- (accessed on 21 June 2012).

31 Moyaux, interview by author, March 2011.

32 "Conversion Planning for Mo-99 Production Facilities from HEU to LEU," working materials, IAEA working group, Vienna, Austria, 24–27 August 2010.

33 From the 1980s until recently, the Belgian producer had obtained only one fresh export of U.S. HEU, in 1988, comprising only 10 kg, sufficient for at most two to three years of production.

34 CERCA, the Company for the Study of Atomic Fuel Creation in France, is an AREVA subsidiary that develops and fabricates nuclear fuel.

35 In 1996, Russia reportedly agreed to export 625 kg HEU for the RHF and Orphée reactors, both in France. In 1998, Russia reportedly agreed to export 400 kg HEU for the FRM-II reactor, in Germany. This latter deal also was said to contain an option for an additional 800 kg HEU, which never was exercised. No further exports from Russia to Europe are known to have occurred, and it appears that neither side wishes to repeat the experience, for reasons that remain somewhat mysterious. Russian HEU first arrived in France in December 1998. See, Ann MacLachlan and Mark Hibbs, "French-Russian HEU Accord Signed; EC Agreed to Russian Prior Consent," *Nuclear Fuel*, 21, 13 (17 June 1996): 1; Mark Hibbs, "Euratom, Again, Accepts Russian Consent for HEU Transfer to FRM-2," *Nuclear Fuel*, 23, 5 (9 March 1998): 1. "Russian HEU delivered to France for use in research reactors," *WISE News Communique*, 5 February 1999.

36 XSNM-3633, U.S. NRC, amended export license application from ILL, RHF, Grenoble, France, received 20 December 2010, in author's possession.

37 XSNM-3633, U.S. NRC, amended export license application, received 21 October 2011, in author's possession.

38 XSNM-3622/01, U.S. NRC, amended export license application, received 30 May 2012, in author's possession.

39 XSNM-3708, U.S. NRC, export license application, received 22 December 2011, in author's possession. Approval granted 2 July 2012, in author's possession.

40 XSNM-3722, U.S. NRC, export license application, received 16 May 2012, in author's possession.

41 "FDA clears Covidien's low-enriched uranium based isotope production," *Health Imaging*, 11 March 2011, http://www.healthimaging.com/index.php?option=com_articles&article=26715 (accessed on 27 September 2012). This article, mainly about South Africa's Mo-99 production, states that "Covidien, based in Dublin, said it is also planning to convert another of its Mo-99 processing facilities, this one in Petten, the Netherlands, to facilitate the use of low-enriched uranium."

# 7 China

## Reactors and Nuclear Propulsion

*Shing-yao (Sandra) Feng*

This chapter provides updates on China's uranium enrichment levels for civilian nuclear activities and military naval propulsion, and the country's efforts to convert research reactors from highly-enriched uranium (HEU) to low-enriched uranium (LEU) fuel. It also examines the internal motivation and external incentives that have influenced China's usage of HEU and LEU for naval propulsion and civilian activities. The existing literature fails to explain what has motivated China since 1995 to move toward conversion or shutdown of HEU-fueled research reactors. Moreover, no previous study has examined the case of China to explore the kinds of incentives and circumstances that can lead governments to phase out HEU use for non-weapons purposes. Understanding China's current policy on uranium enrichment levels for non-weapons uses could also help forecast China's position in nonproliferation negotiations, such as the ongoing talks on the Fissile Material Cut-off Treaty (FMCT), and whether it should be expanded to eliminate HEU for naval propulsion. To help policymakers reduce risks of nuclear terrorism and nuclear proliferation, this chapter develops recommendations on how China and other countries can move towards a comprehensive phase-out of non-weapons use of HEU.

Examining China's decisions regarding enrichment levels can elucidate both China-specific and more general global challenges to the reduction of non-weapons HEU. In Asia, China plays an especially important demonstration role. As countries like Vietnam and Indonesia expand their nuclear programs, they could well follow China's example on enrichment levels. More generally, China has the potential to demonstrate to other governments how each country can find ways to overcome technical, economic, and political obstacles to use of LEU fuel. For example, as the country with the longest experience using LEU for naval propulsion, China can offer insight to three countries that continue to use HEU for naval propulsion: the United States, UK, and Russia. Chinese efforts to convert research reactors that it has exported, and to repatriate HEU fuel from recipient countries, also could be a valuable contribution to international nonproliferation efforts, especially in light of concerns regarding China's export of nuclear technology and reactors.

The key findings of this chapter offer insight on Chinese policy on uranium enrichment levels. First, based on discussions with U.S. and Chinese officials, the chapter reveals that China is currently in the beginning stages of shutting down its miniature neutron source reactor (MNSR) in Shandong, thereby decreasing the remaining number of HEU-fueled research reactors in China from five to four. The chapter also finds that China is taking important steps toward converting Chinese-origin MNSRs domestically and abroad. On 17 September 2010, China signed a multi-year agreement with the U.S. Global Threat Reduction Initiative (GTRI) to construct a Zero-Power Reactor to test LEU fuel for MNSR conversions. The U.S. Department of Energy (DOE) also holds China responsible for the conversion of the five MNSRs that China has exported – to Ghana, Iran, Nigeria, Pakistan, and Syria. Finally, the chapter helps explain China's consistent use of LEU fuel in its nuclear submarines, from the start of its nuclear navy in the late 1950s until today. At least two Chinese experts linked the original decision to Chinese government concerns about the limited amount of HEU available for weapons at the time, and attributed the persistent use of LEU fuel to a type of path-dependency.

The remainder of this chapter is organized as follows. First, it presents a brief overview of China's policy on uranium enrichment levels for non-weapons purposes based on existing literature, followed by an explanation of the methodology used in conducting the research for this chapter. The main body of this report then details findings about China's HEU-fueled reactors, reactor conversion efforts, and China's reported use of LEU for naval propulsion. The chapter concludes with lessons for the United States and the international community about China's use of enriched uranium for non-weapons activities and China's perspectives on the use of HEU in relation to nonproliferation.

## Literature Review

Information in the public domain about China's stockpile of weapons-grade material for military and civilian uses is extremely limited. However, the general consensus is that China's HEU stockpile and small number of HEU-fueled civilian facilities pose a lower proliferation and terrorism threat than those in most other Nuclear Weapons States (NWSs). China was reportedly the first state to use LEU fuel in its nuclear-powered submarines from the beginning, and it did so without international pressure. Today, the United States, UK, and Russia continue to use HEU for naval propulsion, while experts believe that China continues to use LEU. Past studies have not explained the motivations behind China's policies that generally favor LEU over HEU in non-weapons nuclear activities.

### *China's Civilian HEU Policy*

China began producing HEU in 1964 at the Lanzhou gaseous diffusion plant, and in 1975 at the larger Heping plant, the existence of which China refuses to acknowledge.[1] Both enrichment facilities stopped producing HEU by 1987 but continue to produce LEU for civilian uses.[2] Various unofficial statements between 1989 and 1991 reported that China had ceased HEU production for military purposes.[3] In an effort to confirm China's decision to cease high enrichment, Albright and Hinderstein spoke with a Chinese official who confirmed these informal statements but added that the Chinese government will not make an official statement.[4] China remains the only NWS in the Nuclear Non-Proliferation Treaty (NPT) that has not officially announced an end to HEU production for weapons and that has withheld information on its HEU inventory. Although China has been a strong supporter of Fissile Material Cut-off Treaty (FMCT) negotiations, China currently opposes amending the FMCT to require the elimination of previously produced surpluses of fissile material, because that would require China to declare its plutonium and HEU inventory.[5] Some argue that this stance might change if China deployed more survivable nuclear forces.

Hui Zhang reports recent estimates that China has produced a total of 20 ± 4 metric tons (MT) of HEU, and that its current HEU stockpile is 16 ± 4 MT.[6] According to Albright and Hinderstein's calculations, of the 21.5 MT of military and civilian HEU at the end of China's production of HEU, only 500 kg were assigned to civilian use.[7] According to William Potter, the amount of civilian HEU in China is roughly twice that, around 1,000 kg (1 MT), as of 2008.[8] Based on these estimates, there is enough civilian HEU in China to construct at least 10 to 20 nuclear weapons, should the material reach the hands of terrorists.

In 1995, the U.S. DOE and the China Atomic Energy Authority (CAEA) signed a Statement of Intent endorsing cooperation between the U.S. Reduced Enrichment for Research and Test Reactors (RERTR) program and Chinese laboratories.[9] However, China did not officially join the RERTR program. Since the 1995 Statement of Intent, China has engaged in limited cooperation with the RERTR program – and subsequently with the U.S. Global Threat Reduction Initiative (GTRI), after its establishment in 2004 – initiating discussions about the possible conversion of China's HEU-fueled reactors. This engagement has coincided with an increase in China's efforts to convert or shut down its HEU-fueled reactors. In 2007, China converted the HFETR research reactor and the HFETR critical assembly, at the Leshan Nuclear Power Institute of China, and shut down the MNSR-SH in Shanghai.[10]

### *China's Civilian Nuclear Activities*

Based on various sources, four operating research reactors are fueled with HEU in China (see Table 7.1): MNSR-IAE, MNSR-SZ, Zero-Power Fast Critical Reactor (in Chengdu), and China Experimental Fast Reactor (CEFR).[11] Most sources classify the fuel in China's PPR Pulsing Pool UZRH reactor in Chengdu as either HEU or "middle-enriched uranium (MEU)," because it is reported as 20%-enriched, which is just above the threshold for LEU.[12] However, Ray Sollychin of the International Atomic Energy Agency clarifies that China's PPR uses 19.9%-enriched uranium fuel, which makes it an LEU-fueled reactor.[13]

The China Institute of Atomic Energy (CIAE) currently operates the CEFR using 240 kg of 64.4% enriched uranium, but it intends eventually to fuel the fast breeder reactor with uranium-plutonium "mixed-oxide" (MOX) fuel.[14] The

*Table 7.1* China's HEU-fueled Research Reactors, as of 2011

| | *Enrichment* | *HEU (kg)* | *Power (MWt)* | *Conversion Plans* |
|---|---|---|---|---|
| China Experimental Fast Reactor | 64.4% | 240 | 65 | CIAE intends to use MOX fuel in future |
| Zero-Power Fast Critical Reactor | 90% | 41 | 0.00005 | none reported |
| MNSR-IAE | 90% | 1 | 0.027 | CIAE currently working with RERTR |
| MNSR-SZ | 90% | 1 | 0.03 | CIAE currently working with RERTR |

Zero-Power Fast Critical Reactor, a critical assembly used to model such fast reactors, has a core with 40.65 kg of 90%-enriched uranium.[15] There are no reported efforts to convert these two fast-neutron facilities. However, China was able to convert the HFETR Critical Assembly at the Leshan Nuclear Power Institute from 90%-enriched uranium to LEU in April 2007.[16] This indicates that China has had experience in conducting feasibility studies for and converting critical assemblies, which could help it reduce the enrichment levels at CEFR and the Zero-Power Fast Critical Reactor if it chose to. Some Western experts actually advocate decommissioning most critical assemblies entirely and replacing them with computer simulations to model reactors.[17] This could be an effective and inexpensive alternative to converting China's fast critical assembly.

The CIAE also manufactures Miniature Neutron Source Reactors (MNSRs) that are fueled with small amounts of HEU. When the RERTR held a technical meeting on the conversion of MNSRs in Vienna, Austria, in May 2005, there were nine MNSRs in the world: four in China, and the other five exported by China to Ghana, Iran, Nigeria, Pakistan, and Syria.[18] According to the GTRI, China shut down MNSR-SH (Shanghai) in March 2007 and MNSR-SD (Shandong) in December 2010.[19] The CIAE is working with GTRI and the IAEA to convert all seven remaining, Chinese-origin, HEU-fueled MNSRs domestically and abroad. A feasibility study of MNSR conversion to LEU conducted by Jonah, Ibikunle, and Li concluded that MNSRs can be converted to LEU, but the reactor power must be raised to sustain the nominal flux.[20] Jim Matos of Argonne National Laboratory also has conducted a feasibility study for the MNSRs, finding that conversion is feasible with any of five potential LEU fuels – subject to fuel qualification, licensing, and price.[21]

## *China's Submarines and Naval Reactor Fuel*

Nonproliferation experts have called for the FMCT to include a halt to production of HEU for naval propulsion to eliminate a loophole that could enable states to divert naval HEU toward weapons.[22] China is unique among the NWS because its naval propulsion program focused on LEU fuel from the start of research and development in 1958.[23] Hui Zhang notes that the Chinese government did not want the submarine program to compete for HEU with its nuclear-weapons program.[24]

As of 2010, China had an estimated 12 nuclear-powered submarines.[25] The first generation vessels – the Han-class Type 091 attack submarine (SSN), and the Xia-class Type 092 ballistic missile submarine (SSBN) – have suffered technical problems regularly.[26] Furthermore, the Xia and Han classes do not have long-range mission capabilities.[27]

There is no published confirmation that the second-generation vessels – Shang-class Type 093 SSN, and Jin-class Type 094 SSBN – use LEU fuel.[28] Some reports speculate that the 093 SSN might be using an LEU-fueled, high-temperature gas-cooled reactor (HTGR). The HTGR is considered an ideal reactor for submarines because it is compact, high-powered, and quiet. However, experts are skeptical that the 093 uses an HTGR, in light of the substantial technical challenges and the high risk of investing in an advanced but immature technology.[29]

## Methodology

I used secondary sources to obtain background information on China's HEU stockpile and the fuel enrichment level in research and naval reactors. I then cross-referenced secondary and primary sources to determine the most current information available on currently operating HEU-fueled reactors in China.[30] To explore current non-weapon-uses of HEU in China, I conducted in-person interviews with U.S. DOE officials in Beijing, a Chinese arms control expert, a Chinese nonproliferation expert, and a CAEA official. I also conducted phone interviews with officials at the CIAE and the China National Nuclear Corporation (CNNC). This provided both U.S. and Chinese perspectives.

This research aimed to address the following questions:

### *China's Civilian Nuclear Activities*

- What is the most recent progress on converting the MNSRs?
- When will CEFR be fueled with MOX instead of HEU?
- Is it feasible to convert the CEFR to LEU before the loading of MOX fuel?
- Is it technically feasible and are there any efforts to convert the Zero-Power Fast Critical Reactor?

### *China's Submarines and Naval Propulsion*

- What are past and current motivations behind China's use of low-enriched fuel for naval propulsion?
- Are the second-generation submarines powered by LEU-fueled reactors? Are there any plans or consideration of using HEU to fuel naval reactors of a future class of submarines?[31]
- How would the Chinese government feel about an FMCT that included the elimination of HEU for naval propulsion?

### *Cooperation with the U.S. on HEU Minimization*

- What is the status of current U.S.-China cooperation on the conversion of HEU-fueled reactors?
- What expectations does China have in regards to U.S.-China nuclear cooperation? Are there any challenges to cooperation?

There are substantial obstacles to both secondary and field research on this topic. Because China maintains a high-level of secrecy on its HEU stockpile and continues to only offer unofficial statements on HEU production and the uranium enrichment levels for naval propulsion, this chapter cannot present official confirmation on previous reports and estimates. Interviewees refrained from answering questions that were intended to obtain confirmation, especially in regards to naval propulsion. Contacts were more enthusiastic about answering questions related to current conversion projects and perspectives on cooperating with the United States. This chapter cites interviewees based on their affiliated agencies for the purpose of protecting the identity of the individual.

## Findings

### *Civilian Nuclear Activities*

U.S. DOE officials report that GTRI is focused on the conversion of Chinese-origin MNSRs and not the other Chinese HEU-fueled reactors at the moment.[32] There are no plans reported to convert the Zero-Power Fast Critical Reactor or the CEFR. Although neither U.S. officials nor most Chinese officials seemed to be familiar with the Zero-Power Fast Critical Reactor, a CAEA official confirmed that it is still in operation.[33] Both CAEA and CIAE contacts refrained from elaborating on whether China is considering the conversion of this reactor.[34] Jim Matos of the Argonne National Lab stated that although little is known about China's Zero-Power Fast Critical Reactor, the reactor was likely used for criticality experiments in preparation for the construction of the CEFR.[35] Because it is designed as a smaller mock-up of a fast reactor, the neutron leakage is substantially greater than for a larger fast reactor, so higher enrichment is required.[36] Matos concludes that it is not possible to convert the Zero-Power Fast Critical Reactor to LEU but it may be possible to lower the enrichment level to 30 percent.[37]

Officials at U.S. DOE, the CIAE, and CAEA say they do not know when the CEFR will switch to MOX fuel.[38] The CAEA official explained that because spent fuel has to be reprocessed before fabricating MOX fuel for the CEFR, it can be expected that the CEFR will not be able to switch from HEU to MOX in the near future. The official at CIAE says that it cannot lower the enrichment level of the fuel because this would adversely affect the reactor's performance. Matos explained that because the CEFR was designed as a small fast reactor, it faces the same technical obstacle as the Zero-Power Fast Critical Reactor in regards to conversion to LEU. The only way to lower the proliferation risk posed by the CEFR's use of HEU prior to the loading of MOX would be to reduce the enrichment level from 65 percent to around 30 percent.[39]

Neither Chinese nor U.S. officials explained why there are no discussions underway about the lowering of enrichment levels of these reactors.[40] The CEFR is currently fueled with 240 kg of HEU enriched to 64 percent, a substantial amount of fissile material. In addition, the 40.65 kg of 90%-enriched HEU in the Zero-Power Fast Critical Reactor is sufficient for at least one nuclear weapon. These amounts pose a significantly greater threat than the 1 kg of HEU that is in each of the MNSRs. Furthermore, the core in the Zero-Power Fast Critical Reactor is very lightly irradiated, which makes the HEU less self-protecting and thus more vulnerable to theft.

During the 2009 RERTR annual meeting, which happened to be held in Beijing, U.S. DOE and Chinese officials held side meetings to discuss the conversion of China's MNSRs and the construction of a zero-power critical assembly to test LEU fuel for MNSR conversions.[41] Successful tests would enable the fabrication of LEU cores for MNSR conversions. These meetings paved the way for the CIAE to sign a multi-year agreement with the U.S. GTRI on 17 September 2010, to construct a Zero-Power Test Facility (ZPTF). GTRI has agreed to contribute $1.7 million to build the ZPTF, and China has agreed to pay for the cost of the MNSR conversions. An official at the CIAE commented that MNSR conversions do not pose a significant economic challenge to China because of the U.S. funding.[42] The CNNC will build the ZPTF, the CAEA will operate it, the CIAE will provide the LEU fuel, and the United States will continue providing technical expertise and advice. Currently, U.S. DOE is waiting for the CAEA to give final approval to the ZPTF project so construction can begin. The CAEA official says that the Authority is currently conducting calculations on MNSR conversions but does not foresee significant technical challenges.

The CAEA official confirmed that the CIAE has been in the process of shutting down MNSR-SD since December 2010.[43] This leaves two MNSRs left to convert in China: MNSR-IAE and MNSR-SZ. However, U.S. DOE also holds China responsible for the conversion of the five MNSRs that China exported to Ghana, Iran, Nigeria, Pakistan, and Syria. According to the U.S. DOE Office in Beijing, the goal is to build the capacity for China to convert its own MNSRs as well as those that were exported. The U.S. DOE emphasizes the role China plays in converting the exported MNSRs because most of the countries that purchased these reactors do not want to be seen as being told what to do by the United States.[44] An official at the CNNC stated that China's goal is to retain the same performance level after conversion to 1 kg of LEU fuel and then to export additional MNSRs that use LEU.[45] China has already agreed to accept the return of irradiated HEU fuel from the exported MNSRs.[46] A CAEA official confirmed this plan during an interview. All countries that are currently operating a Chinese-exported MNSR, except for Iran, have expressed an interest in conversion if they do not have to bear the cost of conversion.[47]

### *Submarines and Naval Propulsion*

Most interviewees declined to discuss the historical motivation behind China's use of LEU for naval propulsion. A CIAE official stated that anything concerning naval propulsion is a state sovereignty issue and that each country has the right to decide for itself whether to use LEU- or HEU-fueled reactors for naval propulsion.[48] A Chinese arms control expert speculates that one possible motivation is that China may have different performance expectations than the United States or other NWSs for its nuclear-powered submarines.[49] China may not put as much of a premium on building smaller submarines and thus may be more willing to accommodate LEU-fueled naval reactors in larger submarines.

Indeed, the goal of China's first generation submarine construction may have been simply to build a nuclear-powered submarine that worked rather than one that could rival U.S. or Russian submarines. China had the incentive to pursue nuclear-powered submarines in the late 1950s because of ongoing tensions with Taiwan in the context of the Cold War. The inferior performance of Chinese nuclear submarines, compared to U.S. and Russian versions, was unavoidable because China did not possess the economic or technical means to compete. In addition, no other country in East Asia was close to developing nuclear-powered submarines, so even an inferior nuclear submarine would be the region's best. But that may be changing. China's recent push to modernize its military demonstrates the improved economic and technical capacities to pursue better hardware.[50] China also has exhibited more aggression on the seas regarding territorial claims in the region. Vietnam's recent purchase of Russian Kilo-class, diesel-electric submarines will compel China to make sure that it maintains naval superiority.[51] Thus, the argument that China does not have the same priorities as the United States to possess superior military equipment may not hold in the future.

The Chinese arms control expert speculates that another possible reason behind China's use of LEU for naval propulsion is that China, unlike Russia and the United States, never has had a large HEU stockpile. China's reported halt of enriching uranium above 20 percent, two decades ago, limits the amount of HEU available for naval propulsion. China also relies on imports of natural uranium, which makes it more economical to use LEU.[52]

This limited-resource rationale is a plausible explanation for why China decided to design LEU-fueled naval reactors in the late 1950s, when China was simultaneously enriching uranium to develop its nuclear-weapon program. However, China's current economy does not pose any significant obstacles to importing or enriching natural uranium. And even though all of the NWS members of the NPT, except for China, have officially announced an end to HEU production in their respective countries, there is no existing international agreement that would prohibit China from enriching uranium to 20 percent or more. Furthermore, it is unlikely that a version of the FMCT that prohibits HEU production for naval propulsion will be finalized in the near future. Because of this, China could relatively easily produce HEU for naval propulsion, and the international community would find it difficult to argue against this action, especially considering that the United States, UK, and Russia still use HEU-fueled naval reactors. China could argue that any punishment for China's HEU production for naval propulsion would be a discriminatory measure preventing China from pursuing naval technology that other states already have. Thus, the common explanation of a limited HEU stockpile does not explain China's reported continuing use of LEU fuel for naval propulsion.

Chinese officials offered no definitive comments on the fuel enrichment level in China's first two generations of nuclear-powered submarines beyond what is reported in the literature. A CAEA official suggested that the second-generation submarines are likely using LEU fuel because their reactor design is probably based on first-generation reactors; more time would have been required to research, design, and build a completely new propulsion reactor to use HEU fuel.[53] In addition, the CAEA official noted that if the final negotiated FMCT included the elimination of HEU production for naval propulsion, the second-generation submarines would be at a disadvantage if they were designed to use HEU. Accordingly, this official concluded, China has had and continues to have incentive to design and build LEU-fueled nuclear submarines.

The path-dependency logic provided by the CAEA official is the most convincing explanation for China's use of LEU fuel in second-generation naval reactors. It is not because of limited HEU resources or the norms set by the international community, but rather because China's knowledge of submarine design is based on its original LEU-fueled vessels. Although China has the capacity to build HEU-fueled naval reactors, it is easier to build upon past experience than to attempt completely new designs.

Proposals to amend the FMCT to prohibit HEU production for naval propulsion have focused on the United States and other countries that use HEU-fueled naval reactors. But global policymakers should not assume that China fully supports such an amendment to the proposed FMCT, simply because China avoids HEU fuel in its own naval propulsion. China actually would oppose any form of FMCT that requires parties to reveal the size of their existing HEU stockpiles. If amending the FMCT to bar production of HEU for naval propulsion would effectively require China to make public its HEU stockpile, China would oppose that proposal. But the Chinese arms control expert thinks that China publicly would neither support nor oppose such a provision.[54] Considering that other countries including the United States likely would oppose adamantly such an amendment to the FMCT, China could avoid taking any stance and let the others do the work and take the criticism for blocking it.

### *Cooperation with the U.S. on the Reduction of HEU*

The CIAE official stated that there are no formal Chinese policies regarding the use of HEU in civilian nuclear activities.[55] China will continue to act in accordance with the needs and demands of its peaceful nuclear development. However, the official expects that the CIAE will increase cooperation with the United States on the conversion of HEU-fueled reactors, especially after President Hu Jintao's speech at the first Nuclear Security Summit on 13 April 2010, calling for international cooperation and the honoring of nuclear-security obligations. The third of five points outlined in Hu's speech states that "All nations should make full use of existing multilateral and bilateral cooperation mechanisms and channels, share experience on nuclear security, and strengthen information exchange and law enforcement cooperation."[56]

U.S. DOE officials in Beijing noted that Hu's speech was unprecedented.[57] They feel that China had never before emphasized on the world stage the need for bilateral and multilateral cooperation on nuclear security and nonproliferation. Although China had previously produced high-density LEU fuel, and initiated conversion and shut-down activities, such past action was characterized as autonomous rather than in conjunction with international efforts. Indeed, in late 2009, the U.S. DOE in Beijing encountered difficulty in nuclear security cooperation as China underwent bureaucratic changes in its various nuclear governmental bodies.

In addition, U.S.–China nuclear cooperation has often been contingent upon the state of U.S.–China relations. Whenever the United States and China encountered disagreements in foreign affairs, it hindered U.S. DOE nuclear-cooperation efforts in China.[58] For example, U.S. arms sales to Taiwan in January 2010 may have contributed to China's delaying confirmation of President Hu's attendance at the Nuclear Security Summit in April of that year. Some analysts think that Beijing's desire to reduce international perception of China as a threat played a significant role in its ultimate decision to attend the summit.[59] A DOE official in Beijing added that "China significantly stepped up the scope, depth, and number of activities in nonproliferation cooperation after the April 2010 Nuclear Security Summit."[60] The DOE in Beijing has recently experienced good relations with the CIAE and the CAEA, and it reports an increased emphasis by China, in rhetoric and action, on nuclear security.

DOE officials believe that this increase in cooperation since the summit is partly a response to U.S. incentives, including sharing technical information and providing funding such as the $1.7 million contribution to build the ZPTF. Hu's speech also acted as a signal to the Chinese nuclear organizations to pursue bilateral nuclear-security activities, which may have helped push through the September 2010 agreement to construct the ZPTF. A DOE official posits that by increasing cooperative activities, officials at the CIAE and the CAEA can receive more recognition from the central government for following Hu's call to action.[61]

Even with improved coordination on MNSR conversions, certain challenges still confront U.S.–China nonproliferation cooperation. The official at the CNNC argued that it is important for the United States to take the lead and set an example before expecting other countries to convert their own reactors.[62] The official pointed out that the United States is still operating civilian reactors that are fueled with significant amounts of HEU. A Chinese nonproliferation expert in Beijing added that recently expanded U.S.-India nuclear cooperation has had a definite negative impact on China's willingness to cooperate with the United States.[63] For example, the 2009 U.S. nuclear cooperation agreement with India permits that country to enrich uranium, even to weapons-grade. Because China had unofficially announced an end to HEU production, the possibility of India producing HEU causes China to be concerned about its own limited stockpile of this material. China also believes that the U.S.–India agreement, by permitting non-NPT signatory India to engage in activity prohibited to NPT signatories, perverts the intended effects of the treaty.[64]

Such double standards make China feel uncertain about cooperating with the United States on nonproliferation efforts.[65] Indeed, China now has incentive to pursue nuclear commerce with other countries that may undermine international nonproliferation norms on grounds that the United States set the precedent for this kind of behavior. The CNNC signed an agreement in 2010 to provide Pakistan with its third and fourth nuclear power reactors and is considering building an even bigger 1-gigawatt nuclear power plant in that country. China is aware that the United States views this deal as a proliferation concern. It also knows that the sale of nuclear equipment without full-scope safeguards to countries not party to the NPT – whether India or Pakistan – violates the guidelines of the Nuclear Suppliers Group (NSG).[66] This suggests that President Hu's Nuclear Security Summit speech will not translate into universally improved U.S.–China cooperation on nuclear security. Both the United States and China continue to prioritize geopolitics over nonproliferation, at least in South Asia.

Hu also stated in his April 2010 Nuclear Security Summit speech that China seeks to build a nuclear-security demonstration center in China and to play a bigger role in regional nuclear-security cooperation.[67] One of the U.S. DOE officials in Beijing said that Hu's speech and statements by other Chinese officials since the summit suggest that China wants to be a leader in nuclear security in the region.[68] The explanation is not merely that China now recognizes nuclear-security risks, but that its government also wants to be recognized as the regional leader on international issues, a position long enjoyed by Japan despite China's permanent seat on the UN Security Council.

## Summary of Findings

Even though China has moved towards increased cooperation with the United States on the conversion of HEU-fueled reactors, there remain two reactors that have not been part of this dialogue. GTRI has not initiated discussions with its counterparts in China on the possibilities of lowering the proliferation risks posed by the use of HEU in the Zero-Power Fast Critical Reactor and the CEFR. Although it is not possible to convert these reactors to LEU due to greater neutron leakage of small fast reactors, it may be feasible to lower the enrichment levels to about 30 percent.

The 2009 RERTR annual meeting held in Beijing proved fruitful to U.S.–China cooperation on the conversion of MNSRs. Discussion between GTRI and CIAE in the side meetings from that conference led to the agreement to construct a Zero Power Test Facility to develop LEU fuel for all remaining Chinese origin HEU-fueled MNSRs: two in China, and one each in Ghana, Iran, Nigeria, Pakistan, and Syria.

Chinese officials and nonproliferation experts offered several explanations for the military's use of LEU fuel for naval propulsion. One is that China may not seek to have the highest performing nuclear-powered submarines, which would require HEU fuel. Another is that China is constrained by its limited stock of HEU, especially given its unofficial decision to halt HEU production in 1987. The third and most plausible explanation for the reported use of LEU fuel in China's second-generation submarines is based on path dependency. China designed its first generation to use LEU fuel due to its shortage of HEU at the time, and now continues to use LEU because it is easier to build upon the existing design than to develop and test an entirely new, HEU-fueled naval reactor.

U.S. DOE and Chinese officials believe that cooperation between the United States and China on HEU minimization has increased, especially after President Hu Jintao gave a speech on bilateral and multilateral cooperation at the 2010 Nuclear Security Summit. After Hu's speech, the U.S. DOE office in Beijing has experienced improved relations with its civilian nuclear counterparts in China, which may have helped move forward the MNSR conversion efforts. However, some Chinese officials and nonproliferation experts identify critical challenges to U.S.–China cooperation on nuclear security: the continued use of HEU fuel in some U.S. research reactors; and the recent U.S.–India nuclear agreement that adversely affects China's strategic interests.

## Conclusions

This chapter's findings demonstrate that China responds favorably to external incentives that align with its own strategic interests. Some U.S. incentives that helped facilitate progress on MNSR conversion efforts are information sharing and the financial contribution of $1.7 million for the construction of the Zero-Power Test Facility. But China also had an interest in building its reputation as a regional leader on nuclear security issues and as an international stakeholder willing to take on more responsibilities, by advocating greater international cooperation at the first Nuclear Security Summit, in 2010, in Washington, DC.

On the other hand, this chapter also reveals weaknesses in U.S. bilateral cooperation with China on nuclear security. The United States has not engaged China on lowering the enrichment of the uranium fuel in the Zero-Power Fast Critical Reactor and the CEFR prior to the eventual loading of MOX. Each of these facilities holds significantly more HEU than all of the Chinese-origin MNSRs combined. Furthermore, U.S.–India and China–Pakistan nuclear commerce show how U.S–China bilateral relations and nonproliferation goals can easily be undermined by each state's pursuit of narrower national interests.

Regarding naval propulsion, if Chinese second-generation nuclear submarines still do not use HEU fuel, as has been suggested, this would indicate that the Chinese have been able to improve the performance of LEU-fueled naval reactors. Although these submarines do not rival the HEU-fueled submarines of the United States or Russia, the Chinese advances – like those of France – demonstrate to other countries that a nuclear navy can improve performance while avoiding HEU fuel.[69] Continued progress in LEU-fueled submarines could eventually offer a proliferation-resistant alternative for all countries with nuclear navies, enabling a global phase-out of HEU for naval propulsion. Moreover, if path dependency explains why China uses LEU fuel for second-generation submarines and may continue to use LEU fuel for future generations, that is one more reason to insist that any emerging nuclear navies in other states also start with LEU. This would also strengthen the argument for, and increase the feasibility of, an amendment to the proposed FMCT prohibiting HEU production for naval propulsion.

### *Policy Recommendations*

The United States and China should begin talks on lowering the enrichment of the uranium fuel in the Zero-Power Fast Critical Reactor and the CEFR, even if the enrichment can only be reduced to about 30 percent, and regardless of China's plan to eventually switch to MOX fuel. Such a reduction in enrichment would reduce security risks from the fuel, by increasing the amount necessary for a nuclear weapon. It also would support the principle of a global HEU phase-out, which calls for reducing the enrichment of fuel in all reactors worldwide that use HEU.

The fact that U.S. DOE officials could not confirm the status of the Zero-Power Fast Critical Reactor underscores the importance of opening discussions with Chinese officials. The DOE should explore the possibility of conducting feasibility studies with China on using lower enriched uranium for both the Zero-Power Fast Critical Reactor and the CEFR. Chinese counterparts should also explore whether the research and tests conducted using the Zero-Power Fast Critical Reactor could be achieved through computer simulations, which would enable the facility to be closed and decommissioned. DOE and China should not be complacent about the CEFR simply because the use of HEU is foreseen as temporary.

A U.S. DOE official in Beijing observed that one psychological obstacle to conversion is that some governments feel that HEU brings prestige to their country. To counter this sense of prestige attached to HEU, the United States and its European partners should share with China and other states the technology for fabricating the most sophisticated, highest-density LEU fuels, which can enable reactors to essentially match the performance of HEU fuel. The CAEA and U.S. DOE in China think that more information sharing, such as at the 2009 RERTR meeting in Beijing, could eliminate much of the guess-work that confronts facilities when considering conversion. As the United States shares information to help convert Chinese research reactors, China could reciprocate by providing information to facilitate universal adoption of LEU reactors for naval propulsion.

At the moment, the United States is probably the loudest voice internationally calling for HEU minimization. This makes the United States appear to be pressuring other countries to promote U.S. interests, which spurs resistance from some countries that do not want to appear weak. One of the U.S. DOE officials in

Beijing suggested that if other countries, perhaps developing ones, became more vocal about the need to convert or decommission the world's remaining HEU-fueled reactors, some governments might become more receptive.[70]

U.S. nuclear trade with India, and Chinese nuclear trade with Pakistan, have both undermined the NPT and NSG by rewarding states that refuse to sign the NPT or adopt full-scope safeguards. To begin to confront this problem, the United States and China should acknowledge that some of their past actions have created or exploited loopholes. From there, the United States and China could negotiate bilaterally and multilaterally with concerned parties to ensure that future nuclear cooperation does not violate international treaties and agreements.

A lingering question for governments considering conversion of their reactors to LEU fuel, beyond technical challenges, is who should bear the economic burden – the country operating the reactor, the exporter of the reactor, the IAEA, or another third party? Conversion often entails some expense, which can be inhibiting, even if only a marginal cost. To address this concern, the IAEA should develop a comprehensive cost assessment for converting or closing all remaining HEU-fueled reactors in developing countries, and explore ways to fund such conversion or decommissioning.

China already has demonstrated that both internal and external incentives can motivate a state to choose LEU fuel over HEU fuel. But for China to reduce risks from its remaining HEU use, it must work towards lowering the enrichment levels in the CEFR and the Zero-Power Fast Critical Reactor. China also has the opportunity to serve as a model for emerging nuclear naval states by officially announcing its use of LEU fuel for naval propulsion in its first two generations of submarines. This would show that it is possible to establish a nuclear submarine program without HEU. If China continues to build on the steps it already has taken, both autonomously and in cooperation with the United States, it will demonstrate how even a nuclear-weapons state can support a global phase-out of non-weapons HEU.

## Notes

1 David Albright and Corey Hinderstein, "Chinese Military Plutonium and Highly Enriched Uranium Inventories," *Institute for Science and International Security Reports: Global Stocks of Nuclear Explosive Material* (2005): 5, http://isis-online.org/uploads/isis-reports/documents/chinese_military_inventories.pdf (accessed on 15 October 2010).

2 Hui Zhang, "China's Fissile Material Production and Stocks," in *Global Fissile Material Report 2010: Balancing the Books*, (Princeton: Princeton University, 2011): 97, http://belfercenter.ksg.harvard.edu/files/Hui-Zhang-China-Chapter-Global-Fissile-Materials-Report.pdf (accessed on 13 February 2011).

3 Mark Hibbs and Ann MacLachlan, "China Stops Production of Military HEU; All SWU Capacity Now for Civil Use," *Nuclear Fuel* 14.23 (1989): 5. MacLachlan and Hibbs reported that in 1989, the chief of China Nuclear Energy Industry Corp., Zhang Xinduo, stated that China has stopped production of HEU for military purposes.

4 Albright and Hinderstein, "Chinese Military Plutonium and Highly Enriched Uranium Inventories," 5. Albright and Hinderstein report that statements made in 1991 indicated the end of uranium enrichment for weapons.

5 Hui Zhang, "A Discussion of China's Nuclear Transparency Options," paper presented at the INMM 42nd Annual Meeting, Indian Wells, California, 15–19 July 2001, 2–3, http://belfercenter.ksg.harvard.edu/files/inmm01-chinatrans.pdf (accessed on 11 November 2010).

6 Zhang, "China's Fissile Material Production and Stocks," 98.

7 Albright and Hinderstein, "Chinese Military Plutonium and Highly Enriched Uranium Inventories," 5.

8 William C. Potter, "Nuclear Terrorism and the Global Politics of Civilian HEU Elimination," *The Nonproliferation Review* 15,2 (2008): 144.

9 Armando Travelli, "The RERTR Program Status and Progress," paper presented at the 1995 International Meeting on Reduced Enrichment for Research and Test Reactors, Paris, France, 18–21 September 2004, 6, http://www.rertr.anl.gov/PRGM/TRAVEL95.pdf (accessed on 25 October 2010).

10 "GTRI: Reducing Nuclear Threats," National Nuclear Security Administration, last modified 1 February 2009, http://nnsa.energy.gov/mediaroom/factsheets/reducingthreats (accessed on 3 November 2010).

11 Prior to conducting field research 31 December 2010 – 13 January 2011, no publication had mentioned the shutting down of MNSR-SD. The author was informed by Chinese and U.S. officials in Beijing that China was in the process of shutting down MNSR-SD while conducting field research, which occurred before the first published report of its shut-down on 1 February 2011 in "GTRI: Reducing Nuclear Threats."

12 "China: Civilian HEU Reduction and Elimination," Nuclear Threat Initiative, last updated 2008, http://www.nti.org/db/heu/china.html (accessed on 25 October 2010).

13 Ray Sollychin of the International Atomic Energy Agency in discussion with the author, 19 April 2011.

14 Hui Zhang, "China's Fissile Material Production and Stocks," 100.

15 Ole Reistad and Styrkaar Hustveit, "HEU Fuel Cycle Inventories and Progress on Global Minimization," *The Nonproliferation Review*, 15, 2 (July 2008): 266, http://cns.miis.edu/npr/pdfs/152_reistad_appendix2.pdf (accessed on 9 April 2011).

16 "GTRI: Reducing Nuclear Threats," National Nuclear Security Administration, last modified 1 February 2011, http://nnsa.energy.gov/mediaroom/factsheets/reducingthreats (accessed on 9 April 2011).

17 Alexander Glaser and Frank N. von Hippel, "Global Cleanout: Reducing the Threat of HEU-Fueled Nuclear Terrorism," *Arms Control Today* (January/February 2006): 4.

18 Pablo Adelfang, Ira N. Goldman, Adalberto N. Soares, and E. Bradley, "Status and Progress of IAEA Activities on Research Reactor Conversion and Spent Fuel Return Programmes in the Years 2005–2006," paper presented at the 28th International RERTR Meeting, Cape Town, Republic of South Africa, 29 October – 2 November 2006, 2–3, http://www.iaea.org/OurWork/ST/.../RERTR_2006_Adelfang_paper.doc (accessed on 15 February 2011).

19 "GTRI: Reducing Nuclear Threats." The GTRI factsheet page indicates that China confirmed the shutdown of MNSR-SD in December 2010. According to the conversations between the author of this paper and a CAEA official on 10 January 2011, China was still in the process of shutting down MNSR-SD.

20 S. A. Jonah, K. Ibijunle, and Y. Li, "A Feasibility Study of LEU Enrichment Uranium Fuels for MNSR Conversion Using MCNP," *Annals of Nuclear Energy*, 36 (2009): 1286.

21 J. E. Matos and R. M. Lell, "Feasibility Study of Potential LEU Fuels for a Generic MNSR Reactor," paper presented at the 27th International RERTR Meeting, Boston, Massachusetts, 6–10 November 2005, 10, http://www.rertr.anl.gov/RERTR27/Abstracts/S10-2_Matos.html (accessed on 23 November 2010).

22 Chunyan Ma and Frank Von Hippel, "Ending the Production of Highly-Enriched Uranium for Naval Reactors," *Nonproliferation Review* 8 (2001): 86.

23 John Wilson Lewis and Litai Xue, *China's Strategic Seapower: The Politics of Force Modernization in the Nuclear Age* (Stanford: Stanford UP, 1994), p. 24. Lewis and Xue noted that the researchers and policymakers involved in this initial nuclear-powered submarine project (Project 09) decided to use LEU for the submarine fuel right at the beginning of research and development, but they do not discuss the motivations behind this decision.

24 Zhang, "China's Fissile Material Production and Stocks," 101. Zhang cites the following as the source from which he inferred a possible initial reason for the use of LEU in Chinese nuclear submarine propulsion: Li Jue, Lei Rongtian, Li Yi, and Li Yingxiang (eds), *China Today: Nuclear Industry*, (China Social Science Press, Beijing, 1987) (in Chinese), p. 239. Albright and Hinderstein, "Chinese Military Plutonium and Highly Enriched Uranium Inventories," 2. China's first enrichment facility, the Lanzhou gaseous-diffusion plant, did not start producing HEU until 14 January 1964, nine months before China tested its first nuclear explosive. Thus, it is likely that the government wanted to ensure that the nuclear submarine program would not compete with the nascent weapons program for the small amount of weapon-grade material. But this does not explain China's continued reported use of LEU for its second-generation submarines.

25 "China Munitions and Weapons Systems," Federation of American Scientists, last updated 2010, http://www.fas.org/programs/ssp/man/rowwpns/china.html (accessed on 25 October 2010).

26 Lewis and Xue, *China's Strategic Seapower: The Politics of Force Modernization in the Nuclear Age*, 109.

27 Andrew S. Erickson and Lyle J. Goldstein, "China's Future Nuclear Submarine Force: Insights from Chinese Writings," *Naval War College Review* 60,1 (Winter 2007): 65.

28 Lyle Goldstein and William Murray, "Undersea Dragons: China's Maturing Submarine Force," *International Security* 28,4 (Spring, 2004): 171. The second-generation nuclear powered submarines are technologically similar to the Russian Victor III. Potter, "Nuclear Terrorism and the Global Politics of Civilian HEU Elimination," 143.

29 Erickson and Goldstein, "China's Future Nuclear Submarine Force," 67–68.

30 "China: Civilian HEU Reduction and Elimination;" "GTRI: Reducing Nuclear Threats;" and "Nuclear Research Reactors in the World," IAEA, last modified 2010, http://www.iaea.org/worldatom/rrdb/ (accessed on 22 January 2010).

31 This is a concern because China is pursuing long-range ballistic missile submarines, which require a longer core life. Would China opt to use HEU-fueled naval reactors for a new generation of submarines if creating an LEU fuel with a high enough density and burn-up posed too great a challenge?

32 U.S. DOE official in discussion with the author, 6 January 2011. U.S. DOE official in discussion with the author, 11 January 2011.

33 CAEA official in discussion with the author, 10 January 2011. CIAE official in discussion with the author, 4 January 2011. U.S. DOE official in discussion with the author, 6 January 2011. U.S. DOE official in discussion with the author, 11 January 2011.

34 CAEA official in discussion with the author, 10 January 2011.

35 Jim Matos of the Argonne National Lab in discussion with the author, 18 April 2011.

36 Frank von Hippel in discussion with the author, 12 April 2011.

37 Jim Matos of the Argonne National Lab in discussion with the author, 18 April 2011.

38 CAEA official in discussion with the author, 10 January 2011.

39 Jim Matos of the Argonne National Lab in discussion with the author, 18 April 2011.

40 Sharon Sqaussoni, "Looking Back: The 1978 Nuclear Nonproliferation Act." *Arms Control Today* (December 2008): 64. Also see: Edwin S. Lyman, "The Limits of Technical Fixes," in *Nuclear Power and the Spread of Nuclear Weapons: Can We Have One without the Other?*, ed. Paul L. Leventhal et al. (Washington: Brassey's,

2002), pp. 169–170. Talks about the CEFR will inevitably raise U.S. concerns over spent-fuel reprocessing for nuclear energy. The main argument supporting spent-fuel reprocessing, for a closed-fuel cycle, is that it conserves uranium resources and decreases problems regarding spent-fuel storage. However, nonproliferation experts such as Edwin Lyman argue that even a "proliferation-resistant" closed fuel cycle carries proliferation risks and creates false confidence in the fuel cycle. In the late-1970s, the United States formalized its decision to use LEU fuel in the once-through cycle instead of separated plutonium in a closed-fuel cycle, which France, Russia, the U.K., Japan, India, and China have continued to pursue. It may be because the United States is currently re-evaluating the closed-fuel cycle and the possibility of a proliferation-resistant closed fuel cycle that it does not want to make any strong statements yet against China's current reprocessing plans.

41 U.S. DOE official in discussion with the author, January 6, 2011. U.S. DOE official in discussion with the author, 11 January 2011.

42 CIAE official in discussion with the author, 4 January 2011.

43 CAEA official in discussion with the author, 10 January 2011.

44 U.S. DOE official in discussion with the author, 11 January 2011.

45 CNNC official in discussion with the author, 4 January 2011.

46 Adelfang et al., "Status and Progress of IAEA Activities on Research Reactor Conversion and Spent Fuel Return Programmes in the Years 2005–2006," 4.

47 Ibid, 2.

48 CIAE official in discussion with the author, 4 January 2011.

49 Chinese arms control expert in discussion with the author, 5 January 2011.

50 Michael Wines and Edward Wong, "China's Push to Modernize Military is Bearing Fruit," *New York Times*, 5 January 2011, http://www.nytimes.com/2011/01/06/world/asia/06china.html?ref=territorialdisputes (accessed on 14 March 2011).

51 "Vietnam Buys Russian Kilo Class Subs, SU-30 Fighters, Nuke Plant," *Defense Industry Daily*, last modified 31 March 2010, http://www.defenseindustrydaily.com/Vietnam-Reportedly-Set-to-Buy-Russian-Kilo-Class-Subs-05396/ (accessed on 10 March 2011).

52 Chinese arms control expert in discussion with the author, 5 January 2011.

53 CAEA official in discussion with the author, 10 January 2011.

54 Chinese arms control expert in discussion with the author, 5 January 2011.

55 CIAE official in discussion with the author, 4 January 2011.

56 "Text of Chinese President's Speech at Washington Nuclear Security Summit 13 April," *BBC Monitoring: International Reports*, 14 April 2010.

57 U.S. DOE official in discussion with the author, 6 January 2011. U.S. DOE official in discussion with the author, 11 January 2011.

58 Ibid.

59 John Pomfret, "As Nuclear Summit Nears, China Struggles with Role as Rising Power," *The Washington Post*, 29 March 2010.

60 U.S. DOE official in discussion with the author, 6 January 2011.

61 U.S. DOE official in discussion with the author, 6 January 2011.

62 CNNC official in discussion with the author, 4 January 2011.

63 Chinese nonproliferation expert in discussion with the author, 7 January 2011.

64 Jabin T. Jacob, "Indo-US Nuclear Deal: The China Factor," Institute of Peace and Conflict Studies Special Report 14, March 2006, 1, http://www.ipcs.org/pdf_file/issue/1963863503IPCS-Special-Report-14.pdf (accessed on 9 April 2011).

65 Chinese nonproliferation expert in discussion with the author, 7 January 2011.

66 Mark Hibbs, "Pakistan Deal Signals China's Growing Nuclear Assertiveness," Carnegie Endowment for International Peace, 27 April 2010, http://www.carnegieendowment.org/publications/index.cfm?fa=view&id=40685 (accessed on 14 March 2011).

67 "Text of Chinese President's Speech."

68 U.S. DOE official in discussion with the author, 6 January 2011.
69 *The People's Liberation Army Navy: A Modern Navy with Chinese Characteristics* (Suitland, MD: The U.S. Office of Naval Intelligence, 2009): 25, http://www.fas.org/irp/agency/oni/pla-navy.pdf (accessed on 9 April 2011).
70 U.S. DOE official in discussion with the author, 11 January 2011.

# Part IV

# Stubborn Civilian Commerce

# 8 Germany

## The FRM-II Reactor

*Kenneth Dayman*

Going critical on 2 March 2004, the *Forschungsneutronenquelle Heinz Maier-Leibnitz* (FRM-II) in Garching, Germany, outside Munich, became the world's first (non-miniature) research reactor in more than a decade to be fueled by highly enriched uranium (HEU), a nuclear weapons-grade fissile material.[1] As such, the reactor represented a step backwards in the global movement to reduce nuclear-security risks by eliminating the unnecessary usage of HEU for non-weapons purposes. The new, HEU-fueled German reactor threatened to undermine that emerging norm and ignite an international debate over equal access to nuclear technology, impeding efforts to discourage HEU usage, just as more countries were developing nuclear programs.

Annually, the FRM-II now uses four fuel elements, each containing 8.1 kg of 93%-enriched HEU, in 60-day cycles.[2] The reactor thus uses 32.4 kg of weapons-grade uranium per year. This is considerably more than a significant quantity (SQ) of HEU, which requires only 25 kg of U-235, as defined by the International Atomic Energy Agency (IAEA).[3] Even after discharge, a year's worth of spent fuel still contains roughly 28 kg of 88%-enriched uranium.[4]

Currently, the United States, Europe, and Russia are developing very high-density LEU fuel, aiming to convert all HEU-fueled high performance research reactors (HPRRs). All five U.S. HPRRs are currently slated for conversion by 2021, under the schedule of DOE's Global Threat Reduction Initiative (GTRI); two European HPPRs have pledged to convert; and Russia in 2012 will complete conversion studies for six of its research reactors, although not yet its highest-performance reactors.[5]

The FRM-II's final license originally required the reactor to convert by the end of 2010 to fuel of no more than 50 percent enrichment, which is still above the 20 percent threshold defining HEU. Due to delays in the development of an appropriate fuel, the deadline was postponed to 2018.[6] Even if the reactor were converted by that time to 50%-enriched fuel, it could soon find itself as the only HEU-fueled research reactor left in "model-setting countries."[7] But the conversion deadline may again be bypassed if the new, high-density fuel is still not available by that time. The FRM-II is also slated to begin irradiating targets for the production of the medical isotope molybdenum-99 (Mo-99), but the enrichment of the targets will be determined by the producers, who currently use HEU.

This chapter next reviews the literature on the potential for conversion of the FRM-II. It then outlines my research methodology. The heart of the chapter presents my detailed research findings. The chapter concludes with the ramifications of these findings and recommendations for the various involved parties to encourage continued progress towards conversion. Specifically, I identify several sources of leverage to encourage more timely conversion of the FRM-II. The leverage is provided by the front-end and back-end of the FRM-II's fuel cycle, and by ongoing U.S. efforts to convert domestic, high-performance research reactors.

## Literature Review

The published material regarding FRM-II and its potential conversion is diverse. Primary sources include official documents, such as government agreements and licenses. Secondary sources include peer-reviewed journal articles, conference presentations and proceedings, dissertations, newspaper articles, and press releases.

A great deal is known about the design of FRM-II, as it has been studied by various groups, most notably the Reduced Enrichment for Research and Test Reactors (RERTR) group at Argonne National Laboratory (ANL) in the mid-1990s, when it proposed alternative designs for the reactor to use LEU fuel.[8] In response to concerns from FRM-II operators about the impact that conversion to LEU fuel would have on the scientific utility of the reactor, further studies were carried out to examine various performance characteristics of the reactor in greater detail under different conversion scenarios.[9] The different designs were simulated using Monte-Carlo neutron transport simulations (e.g., with MCNP) in order to ascertain the effects on significant parameters of the reactor's performance as a neutron source. These include the magnitude and spatial distribution of the neutron flux, the neutron spectrum throughout the moderator tank, where entrances to the beam ports that feed the neutron guides and experiments are located, and the heat generated by fission, especially as it pertains to the heating of the cold and hot neutron sources also located in the moderator tank. The methods of these simulations and the results are well documented in the literature.[10]

In 1996, construction of the facility began, making a redesign less viable, especially if changes were required to the size of the core; however, critics already had begun to advocate for conversion. In response to a changing political atmosphere in Germany following the 1998 elections, the *Bundesministerium für Bildung und Forschung* (German Federal Ministry of Education and Research, BMBF) established an expert commission in 1999 to study the conversion of FRM-II.[11] Out of this commission came several conversion options.

In a 2003 agreement between the BMBF and the Bavarian government, it was decided to start up the reactor using 93%-enriched HEU as originally planned and subsequently convert the reactor to fuel of enrichment no greater than 50 percent ("middle-enriched" uranium, or MEU)[12] by the end of 2010. This requirement was contingent on the successful development and qualification of an appropriate fuel of adequate density to allow for conversion, as specified in the reactor's license, issued by the *Bundesministerium für Umwelt, Naturschutz, und Reaktorsicherheit* (Federal Ministry for Environment, Nature Conservation, and Nuclear Safety, BMU).[13] Thus, FRM-II was allowed to start up using HEU fuel and went critical for the first time on 2 March 2004, achieving its designed full power of 20 MW on 24 August 2004. In the meantime, research and development into new high density fuels has continued, most notably at ANL and Idaho National Laboratory (INL).

In October 2009, representatives from FRM-II presented a paper at the 2009 RERTR meeting exploring the feasibility of Mo-99 production at their facility.[14] In February 2011, a new thimble was installed in the moderator tank of the reactor in order to allow for the irradiation of uranium targets for the production of Mo-99,[15] which is slated to commence in 2014.[16] The specifics of the prospective isotope production are unclear – what form the targets will take, their enrichment, and where and by whom the chemical separations and processing will be performed.

On 26 July 2010, *Die Sueddeutsche Zeitung* published a story in which Bavarian Minister-President Seehofer said that FRM-II would not be converted to MEU by the end of the year and HEU would continue to be used in the reactor until 2018.[17] Russian-supplied HEU, which was fabricated into fuel elements by the French company CERCA, has been used most recently to fuel the reactor. But that current supply is set to run out sometime in 2015 or 2016, and an operator could not identify an additional source, when queried at an international conference in January 2012.[18] It is unclear whether the operators have serious intentions of reducing the fuel enrichment, or what may be expected in 2016 and 2018.[19]

One difficulty that arises in examining the literature on projected performance under various conversion scenarios is deciding how to compare results from different researchers and determine the most credible results. This is due not to a lack of expertise on the part of the different parties who have studied the FRM-II, but differences in assumptions or constraints applied in different studies, and differences in the models generated and used. Resolving these differences and deciding which assessments of feasibility are most reasonable is difficult, as these simulations are very complex and time-intensive, making a comparative evaluation cumbersome. Moreover, the interpretation of the policy implications of these simulations is not entirely objective. For example if a simulation determines objectively that conversion will reduce thermal neutron flux by 10 percent, a subjective assessment must be made as to whether that penalty disqualifies the proposed conversion as "unfeasible."

## Methodology

My research questions were as follows:

- Why was it decided that the FRM-II would not be converted to MEU by the end of 2010?
- With this precedent in mind, should we expect a conversion to take place in 2018?

- Where will the HEU fuel come from to supply the reactor until 2018?
- Is a conversion to MEU still intended?
- Will the targets for proposed Mo-99 production be HEU or LEU?
- What are the plans for performing the chemical separation of Mo-99 from the targets?
- What are possible sources of leverage or incentives that may be employed by interested parties to facilitate progress on conversion of FRM-II to LEU, or at least to lower enriched fuel?

In order to answer these questions, contact was sought with German government officials, FRM-II scientists and operators, U.S. scientists at Department of Energy national laboratories involved with high-density fuel development, interested nonproliferation advocates, and scientists who have studied conversion options for the FRM-II. This primary research was conducted via email, phone, and Skype interviews, and travel to various cities in Germany for in-person interviews.

One major difficulty that arose in the field research was the sensitivity of the subject and German government officials' general unwillingness to discuss it. Another obstacle was that at the moment, while potentially suitable LEU fuel is still being developed and the FRM-II is operated using a stock of HEU fuel, there is little political debate or discussion about converting the reactor.

Contact was made and interviews conducted with three independent experts on the feasibility of converting the FRM-II: Dr. Alexander Glaser at Princeton University, whose dissertation studied conversion options for the FRM-II and their effects on reactor performance;[20] Dr. Matthias Englert, who did further study of conversion scenarios while at the Technical University – Darmstadt;[21] and their advisor at TU – Darmstadt, Dr. Wolfgang Liebert. I visited the FRM-II and contacted its scientific director, Dr. Winfried Petry. I also interviewed a leading German NGO official who has worked on the issue for fifteen years, Dr. Annette Schaper, at the *Hessische Stiftung Friedens und Konfliktforschung* (Peace Research Institute Frankfurt, HSFK), a public policy think-tank located in Frankfurt, Germany. Another interview was conducted with a representative from the BMBF, Stefan Kern, head of the basic science research division. Finally, contact was made with the office of Dr. Norbert Röttgen, the Federal Minister for Environment, Nature Conservation, and Nuclear Safety.

A final obstacle to this research was that of language. I have conversance with German but faced some hurdles in communicating with interview partners and their associates. Reading documents written in German, especially those of a technical nature, and parsing the large volume of technical reporting, such as the dissertations of Glaser and Englert, was also a challenge.

## Findings

The FRM-II is currently fueled by 93%-enriched HEU in the form of $U_3Si_2$ that has a relatively high density of 3 gU/cc. Ironically, the $U_3Si_2$ that enables this high density was originally developed by the RERTR program to produce LEU fuels of sufficient density to replace HEU fuel in research reactors. But since the FRM-II adapted this innovation to increase the density of its HEU fuel, even higher density fuel than currently available would be needed to convert the reactor to LEU. Research into new high-density fuels using alloys of uranium and molybdenum (UMo) is currently under way. There are two varieties of these fuels: dispersion and monolithic.

Dispersion fuels are made by dispersing the fissile material (UMo) in an inert matrix, such as aluminum. This is typically done by generating both materials as powders, which are then mixed and hardened into a solid. Fuels of this type can achieve fuel loadings of 6-8 gU/cc. In the case of FRM-II, this is not sufficiently dense to allow for conversion to LEU (or even MEU).

Monolithic fuels, however, can be made with densities of up to 16 gU/cc and thus enable conversion of HPRRs already using high-density HEU fuel to LEU fuel in nearly all cases.[22] Monolithic fuels are produced by first manufacturing a foil of the fuel meat, comprising some kind of uranium-molybdenum mixture, such as "U10Mo," meaning a mixture of uranium and 10 percent molybdenum. The foil is then sandwiched in between plates of cladding material that are subsequently bonded in some way, such as hot isostatic pressing (HIP) or friction bonding (FB). Unfortunately, some complications have arisen during the irradiation testing of sample fuel plates, with potential implications for FRM-II, as detailed below.

Figure 8.1 is a diagram of the FRM-II's single fuel element. The control rod and coolant ($H_2O$) pass through the center hole ($|r| < \sim 5$ cm), and the element is surrounded by $D_2O$ in the moderator tank when installed (see Figure 8.2). The fuel element includes 113 involute-shaped fuel plates, which are placed in a parallel configuration in the active part of the element ($\sim 5$ cm $< |r| < \sim 12$ cm). The gaps between the individual fuel plates are important for coolant/moderator flow.

The problem with the proposed conversion to monolithic UMo LEU is that, at high fission densities, test plates of this fuel have exhibited swelling, which in the FRM-II could lead to a loss of coolant flow around parts of the fuel element. This renders the monolithic LEU, at least in its current state of development, unsuitable for the FRM-II. Research into solving this problem is currently underway at INL.[23]

Officials at the Technical University – Munich (TUM), home of the FRM-II, report that since 2004 they also have performed a variety of work investigating dispersion and monolithic uranium-molybdenum fuels for the conversion of the reactor. Manufacturing techniques, behavior under high radiation fields, and methods for qualification of these fuels all have been investigated.[24] Numerous papers have been published and presented at conferences detailing these efforts.

Regarding potential conversion of the FRM-II, Table 8.1 shows some parameters and projected outcomes of different fuel types. This data is from an apparently unpublished report produced by scientists at FRM-II to explain some of the ramifications that they expect from a change of fuel. The date of this report is unknown. But since it considers the fuel-cycle length to be the current 60 days, as opposed to the 52 days originally used, it can be presumed that the report was completed no earlier than 2007.

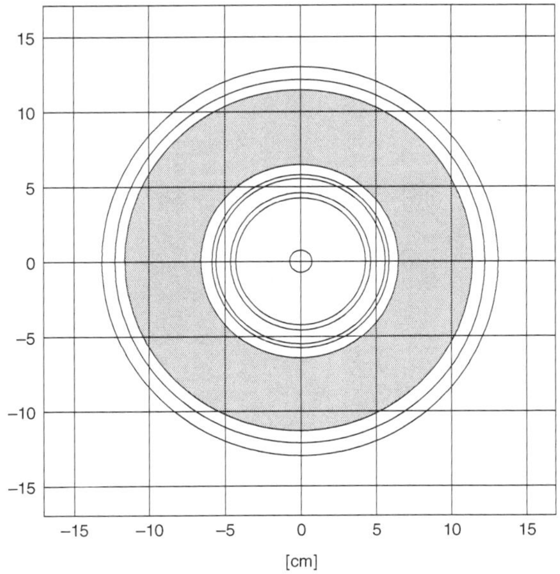


*Figure 8.1* Diagram of the FRM-II Fuel Element

Note: 113 involute-shaped fuel plates are placed inside an annulus configuration with small gaps between the individual plates to allow for coolant/moderator flow.

Source: Alexander Glaser, "The Conversion of Research Reactors to Low-Enriched Fuel and the Case of the FRM-II," *Science and Global Security* (November 2002): 61.

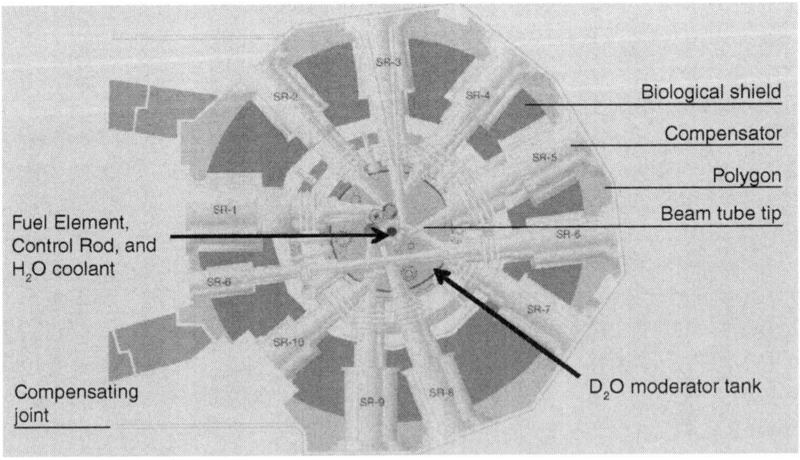


*Figure 8.2* Diagram of the FRM-II Moderator Tank

Note: Shown is the fuel element inside the moderator (D2O) tank, which also houses the entrances to all the beam ports that provide neutrons to the neutron guides.

Source: Adapted from W. Gläser,"New Neutron Source FRM II," *Appl. Phys. A74 [Suppl.]*, S23–S29, 2002.

*Table 8.1* Estimated Effects of Different Fuel Types and Enrichments at FRM-II

| *Enrichment [%]*<br>*Fuel* | *93*<br>*$U_3Si_2$* | *50*<br>*$U_8Mo$ (dispersion)* | *34*<br>*$U_8Mo$ (monolithic)* | *20*<br>*Not Possible* |
|---|---|---|---|---|
| Fuel Density [g/cc] | 3 | 8 | 15 | 40.5 |
| Neutron (Flux) Loss [%] | 0 | 8 | >10 | >>10 |
| Total Uranium Mass [kg] | 8 | 21.4 | 41 | 110 |
| Mass of $^{235}U$ [kg] | 7.5 | 10.7 | 14 | 22 |
| Mass of Pu Produced after 60 day cycle [g] | 13.8 | 87.7 | 137.3 | 253.8 |
| Fissioned $^{235}U$ [%] | 20 | 14 | 10.7 | 6.8 |
| Factor Increase in Residual Radiation after 500 Years of Decay | – | 6.3 | 10 | ~20 |

Source: Unpublished report by the FRM-II staff at Technical University – Munich.

Officials at TUM estimate that conversion of the FRM-II to a fuel of 20%-enrichment (the threshold for LEU) would require a fuel density of 40.5 gU/cc, which is physically impossible because it exceeds the density of pure uranium metal. It is unclear what assumptions were made for these calculations. Scientists at ANL and TU-Darmstadt have presented alternative LEU conversion scenarios, based on realistic fuel densities, that would yield reactor performance comparable to what is currently achieved. The ANL and other researchers did allow for a change in the fuel-element geometry – suggesting an increase in the active fuel height by 10–12 cm (14–17 percent), and an increase in power from 20 to 22 MW (10 percent) – which would reduce the flux loss from using LEU to less than 10 percent.[25] In all conversion scenarios considered, losses in neutron flux are expected, but the significance of these losses is disputed.[26]

Because high-density UMo fuel has not yet been qualified or licensed, it was not possible to convert the FRM-II to MEU by the 31 December 2010 deadline set by the BMBF-Bavarian agreement and the BMU license.[27] These documents do include a caveat that conversion necessitates appropriate fuel to be licensed and ready for use, and some observers suggest that the conversion requirement is thus only enforceable once acceptable fuel has been developed.[28] A new agreement between the BMBF and Bavarian government has been drafted that would allow the reactor to run until 2018, after which the reactor would have to be converted to fuel of enrichment no greater than 50 percent U-235, assuming such fuel is available. The agreement also would require the FRM-II operator in the meantime to pursue development of higher-density, lower enrichment fuel. Reportedly, officials in Bavaria have pushed for loosening some of these provisions in the agreement, which in 2011 was pending approval from the BMU.[29]

In light of the failure to meet the 2010 deadline, it is natural to wonder what can be expected in 2018. If an acceptable fuel is not yet licensed, conversion will not happen by that deadline and the operators likely will seek to continue operation using HEU. Considering the cost of the reactor, the cost of the associated experimental instrumentation, and the volume of work performed by a variety of groups from across Europe and the rest of the world, neither the government nor the operators are interested in shutting down the reactor if the new conversion deadline cannot be met.[30] The operators say they are still committed to conversion to 50 percent enrichment or less, and would like to go to the lowest enrichment reasonably possible. They estimate that this will be somewhere between 34 and 50 percent.[31] As Wolfgang Liebert notes, if all other international parties continue to observe the norm of phasing out civilian HEU use, especially with regards to converting research reactors to fuel enrichments of less than 20 percent, then Germany is more likely to reduce the enrichment of the FRM-II fuel too.[32]

The ultimate enrichment of the FRM-II fuel could have serious implications both for the security risk posed by the facility and for sustaining the global norm of HEU minimization. Figure 8.3 shows the critical mass of a reflected sphere of uranium metal versus the enrichment of the uranium. As expected, assemblies of lower enrichments require a higher mass to become critical, meaning more material is needed to develop a weapon. As can be seen, based on one set of assumptions, approximately nine times more material is needed to form a critical configuration using 20%-enriched uranium (the LEU threshold) compared to the 93%-enriched uranium (HEU) currently used in the reactor. Additionally, as noted, the FRM-II's spent fuel remains highly enriched, and a year's worth contains approximately one significant quantity of HEU, by IAEA standards. If the uranium were separated from the spent fuel assemblies, it would be approximately 88%-enriched and would have a critical mass very similar to 93%-enriched uranium.

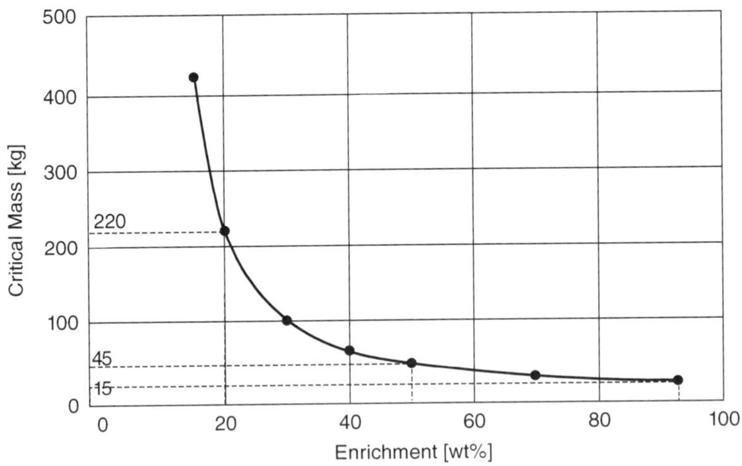


*Figure 8.3* Reducing Enrichment Increases Amount Required for a Weapon

Note: The critical mass of a 10-cm beryllium-reflected sphere as a function of the U-235 enrichment. The critical mass for 93, 50, and 20 percent enrichments are approximately 15, 45, and 220 kg, respectively.

Source: Adapted from Alexander Glaser, "The Conversion of Research Reactors to Low-Enriched Fuel and the Case of the FRM-II," *Science and Global Security* (Nov. 2002).

Because all deadlines and timetables for conversion of the FRM-II are contingent on acceptable fuel being developed, this raises the question of what qualifies as acceptable fuel. Based on interviews with German officials, it seems there is no agreed-upon definition for acceptability. Any reduction in enrichment would have some impact on the reactor's performance and/or cost. For example, conversion to LEU fuel would sacrifice some aspects of neutron flux unless changes were made to the core geometry and placement of the beam tubes within the moderator tank (see Figure 8.2).[33] Such changes to the geometry of the element or the arrangement of objects in the moderator tank would be quite expensive, time intensive, and difficult due to the highly radioactive materials that have been activated during past operation of the reactor.

Conversion to LEU without these changes would undeniably degrade performance, so it is useful to consider whether such losses would be acceptable. Several parameters are relevant in comparing the current FRM-II to a converted version or to other HPRRs. The most relevant of these is the average thermal neutron flux at the entrance to the beam tubes. This is the flow rate of thermalized neutrons incident on the beam tubes, thereby affecting the number of neutrons provided to each experimental instrument over time. Any loss in the flow rate of neutrons per experiment would necessitate longer measurement times, imposing an economic penalty. Another important parameter is the neutron spectrum. For many of the experiments performed at FRM-II, fast neutrons equate to noise in the experimental results, so any change to the energy distribution of the neutron flux affects the quality of the analytic results, imposing a technical penalty. Both types of penalties – economic and technical – could reduce the attractiveness of the facility to private and government funders.

But several other parameters, commonly cited by opponents of conversion, should *not* have a strong bearing on determining whether lower enriched fuel is "acceptable." One of these is the plutonium content of the spent fuel upon discharge, allegedly a proliferation concern. While true that lower-enriched fuel will have a greater concentration of transuranic nuclides including plutonium upon discharge, this amount is miniscule both in terms of proliferation risk and in comparison to the quantity of plutonium in the spent fuel of commercial nuclear power plants, which is orders of magnitude greater.[34] Table 8.1 reports that converting the FRM-II from HEU to LEU fuel would increase the plutonium in each spent fuel element by a factor of 18. However, the total amount of plutonium produced in an LEU fuel element would be only 253.8 g, barely 3 percent of a significant quantity of plutonium, defined as 8 kg by the IAEA.[35] Accordingly, more than 31 spent LEU fuel elements (eight years' worth) would need to be chemically reprocessed, a non-trivial task, to separate out a single SQ of plutonium for weapons purposes.

By contrast, the remaining U-235 in a single spent HEU element still comprises nearly one-fourth of a significant quantity, so that one year's worth of fresh or spent HEU fuel is equivalent to about one SQ. Moreover, unlike plutonium, fresh HEU can be used to make gun-type weapons, which are much easier to construct than implosion designs. This is due to HEU's lower spontaneous fission rate that permits less sophisticated assemblies of a critical mass to achieve impressive weapon yields. In short, the uranium in a current HEU fuel element of the FRM-II represents a much greater proliferation danger than the plutonium in a prospective spent LEU fuel element.

Accordingly, the "acceptability" of higher-density, lower-enriched fuel for the FRM-II, once successfully qualified, should be based on only two factors: performance and cost. But, as noted, this assessment remains somewhat subjective. Some might argue that any cost or performance penalty from conversion is unacceptable, and thus no lower-enriched fuel would ever be acceptable, leaving the 2018 deadline unable to be enforced.

German government officials do not appear ready to discuss these issues. They declined to discuss them in interviews on grounds that the policy decisions were settled years ago. But one German non-governmental critic, Annette Schaper, argues that those earlier policy decisions were made prematurely, without the requisite technical expertise. The German government did organize an expert commission in 1999, but by the time its results were reported, the construction of the facility was too far along for the panel's findings to be adequately considered and have any real impact, she says.[36]

Revisiting the question is difficult in part because of Germany's federal system of government and division of responsibility between government ministries. For example, the only part of the federal government that has serious interest in international relations, and accordingly may be concerned that the use of HEU fuel could undermine nonproliferation norms, is the *Auswärtiges Amt* (Foreign Office, AA), which does not have a position of oversight of the FRM-II.[37] Moreover, the state of Bavaria, where the TUM is located, has long supported FRM-II officials opposed to reducing the fuel enrichment, and German states have significant autonomy under federalism.

With the question of conversion seemingly moot until the successful development of higher-density fuel, the FRM-II may continue running on HEU, but the question of fuel supply arises. The facility's current HEU fuel inventory is estimated to run out sometime in 2015 or 2016.[38] Under the Schumer Amendment to the Energy Policy Act of 1992, the United States will not export HEU to Germany for use as reactor fuel.[39] So if the reactor is to continue operating, HEU will need to come from another country with a stockpile of such material, making Russia the leading if not sole candidate. In a 2011 interview, one German non-governmental expert did not seem concerned about the reactor being able to acquire fresh HEU fuel, deeming it extremely unlikely that European nuclear officials would allow one of their premier research facilities to go dormant due to lack of fuel.[40] But in January 2012, at an international conference, an operator of the FRM-II expressed greater concern, noting that to avoid an interruption in operation he would have to contract to fabricate new fuel by about 2014, but he lacked a source of HEU to do that. The operator expressed hope that the United States eventually would supply HEU, but U.S. officials at the event offered no support for that hope.[41]

With regards to Mo-99 production, FRM-II will be used strictly as an irradiation facility for targets, the enrichment of which will be determined by the actual producer, who will also perform the requisite chemical separations and purifications.[42] It was not revealed who this producer-partner is. The most likely candidates are Europe's two commercial producers of Mo-99: Covidien in the Netherlands, and IRE in Belgium. (For further details, see this book's chapter by Alexander Fay.)

## Summary of Findings

The FRM-II in Garching is required by law to convert to the lowest feasible enrichment by the end of 2018, assuming that acceptable fuel is available by that time. Originally, this conversion was supposed to take place by the end of 2010, but excessive swelling of the monolithic UMo fuel plates currently under development at INL and elsewhere prevented this possibility, and an extension was granted. If an acceptable fuel is not available by the new conversion deadline, the reactor will aim to continue operating on HEU fuel, given the economic investment and the facility's role as a neutron source for a wide array of scientific groups from around the world. It is unclear where fresh HEU would be obtained to fuel the reactor after the current fuel supply is exhausted, estimated to occur in 2015–2016. The FRM-II will be used as an irradiation facility for production of Mo-99, scheduled to begin in 2014, but the enrichment of the target will be left up to the producer who will also perform the requisite chemistry.

## Conclusions and Recommendations

In the context of the global effort to phase out HEU for non-weapons purposes, the FRM-II represents a serious departure and thereby threatens the momentum of that movement. Prospective conversion of the FRM-II to lower-enriched fuel, and the timeframe of such conversion, are dependent on the development, qualification, and licensing of high-density UMo fuel. If this new fuel is not available by the 2018 deadline, the reactor will aim to continue operating as it has on fuel of 93%-enriched HEU, if a sufficient amount can be acquired. Indeed, the current fuel supply is expected to be exhausted no later than 2016.

Further research into the fabrication of very high-density fuels, namely monolithic UMo fuel, must be pursued to optimally reduce the fuel enrichment of several HPRRs, including the FRM-II. Besides an increase in funding, there does not appear to be any way immediately to accelerate this process. Research into the different groups studying monolithic fuels and the extent of their cooperation should be conducted to identify any possible obstacles to progress that could be overcome by better coordination.

Sources of leverage or influence to encourage conversion lie mostly with the United States and Europe setting an example by pursuing the conversion of other HPRRs with vigor. Setting deadlines and timetables for this process would increase moral bargaining power in any negotiations with Germany. The continued conversion of reactors around the world from HEU to LEU would undoubtedly increase pressure on FRM-II to convert too.

Other sources of leverage may include the front-end (fuel supply) and back-end (waste disposal) parts of the FRM-II's fuel cycle. Germany's need for HEU to fuel the reactor from 2016 to 2018, and perhaps beyond, provides the United States potential leverage to encourage more timely conversion. However, exercising this leverage might require the United States to amend its current nonproliferation law to permit temporary export of HEU for use as fuel in the FRM-II.

The FRM-II also has the back-end problem of waste disposal, which is especially challenging with HEU fuel. The uranium in the spent fuel remains approximately 88%-enriched, and the spent fuel also contains additional fissile material produced through the activation and decay of fertile nuclides. This poses a significant risk of criticality that further complicates waste disposition beyond the issues that confront spent LEU fuel from research or power reactors.[43] Even after a conversion to MEU or LEU, the FRM-II's inventory of spent HEU fuel would need to be disposed of with technology yet to be developed. Accordingly, offers to share in the capital and manpower cost to develop this technology, or other burdens of disposal, could be used as a bargaining chip in any discussion regarding conversion.

From a nonproliferation and security standpoint, converting the FRM-II to LEU fuel would be far superior to MEU fuel. As noted in Figure 8.3, converting from 93 percent to 50 percent enrichment would only triple the critical mass, providing limited benefit. Moreover, conversion to 50 percent MEU could undermine the international nonproliferation norm that only LEU is acceptable. Some experts argue that conversion will only occur once at the FRM-II, so it may be more desirable to be patient and possibly delay conversion in order to achieve the lowest possible enrichment.[44] Accordingly, any potential agreement between the United States and Germany should make a commitment to convert to the lowest enrichment possible – ideally LEU – in order to obtain the maximum nonproliferation benefit from conversion of the FRM-II.

To avoid repeating the poor timing of the 1999 expert commission, which came too late to have any serious impact on policy decisions, now is the ideal time to start real policy discussions and debates – in advance of the FRM-II exhausting its current fuel supply by 2016, or reaching its deadline for conversion in 2018. German NGOs could initiate a campaign to raise German public awareness of the detrimental effect on international nonproliferation efforts of continued HEU use at the FRM-II. U.S. officials could encourage the German government to begin exploring options as soon as possible for reducing enrichment of the fuel at the FRM-II.[45] Most important, the German government must consider what constitutes "acceptable" lower-enriched fuel – i.e., what technical and economic penalties it is willing to accept at the FRM-II in order to comply with the international norm of phasing out HEU as fuel in research reactors.

## Notes

1 Preceding that was China's MJTR in 1991, http://www-naweb.iaea.org/napc/physics/research_reactors/database/rr%20data%20base/datasets/report/China,%20People's%20Republic%20of%20%20Research%20Reactor%20Details%20-%20MJTR.htm (accessed on 27 September 2012). China also subsequently produced miniature neutron source reactors (MNSRs), but those each contain only 1 kg of HEU in a lifetime core, compared to the FRM-II that was projected to use more than a ton (1000 kg) of HEU over its lifetime.

2 During its first few years of operation, the FRM-II followed its originally planned fuel cycle comprising five fuel elements annually, with a 52-day cycle, requiring a total of 40.5 kg HEU in fresh fuel annually. In 2007, this was changed to four 60-day cycles. (Interviews with Winfried Petry, 22 February 2011 and 7 May 2011.)

3 The amount of a material defined to be a significant quantity is related to the amount of the material required to produce a weapon, i.e., the critical mass of a material in some geometry. The critical mass is highly dependent on the geometry and, thus, the amount of fissile material required to make a weapon depends on the details of the weapon being produced. For the purposes of defining significant quantities, reflected spheres are typically considered. See International Atomic Energy Agency, *IAEA Safeguards Glossary*, IAEA/SG/INF/1, 1980.

4 Alexander Glaser, "The Conversion of Research Reactors to Low-Enriched Fuel and the Case of the FRM-II," *Science and Global Security* 9 (November 2002): 74. See also, Alexander Glaser, "Neutronics Calculations Relevant to the Conversion of Research Reactors to Low-Enriched Fuel," Ph.D. diss., Technischen Universität Darmstadt, 2005, p. 165. These sources estimate that the uranium in the spent fuel would have an enrichment of about 88.5 percent, after the FRM-II's original 52-day fuel cycle. This enrichment might be slightly lower after the 60-day fuel cycle now being used.

5 See chapters in this book by Margarita Jimenez and Braden Civens.

6 The enrichment requirement is the same, requiring conversion to fuel of no more than 50 percent U-235, provided an appropriate fuel is qualified and ready for use.

7 R. Kreibich, and Wolfgang Liebert, Vereinigung Deutscher Wissenschaftler to Ministerin Edelgard Bulmahn, 24 Februrary 2003, Personal Files of Wolfgang Liebert.

8 S.C. Mo, N.A. Hanan, and J.E. Matos, "Comparison of the FRM-II HEU Design with an Alternative LEU Design," paper presented at the 1995 International Meeting on Reduced Enrichment for Research and Test Reactors, Paris, France, 18–21 September 1994. N.A. Hanan, S.C. Mo, R.S. Smith, and J.E. Matos, "An Alternative LEU Design for the FRM-II," RERTR Program, Argonne National Laboratory, 1996.

9 Characteristics such as heating of the cold neutron source, neutron fluxes at the experimental facilities, and reactor transients are examined in N.A. Hanan and J.E. Matos, "Fluxes at Experiment Facilities in HEU and LEU Designs for the FRM-II," paper presented at the 1997 International Meeting on Reduced Enrichment for Research and Test Reactors, Jackson Hole, Wyoming, USA, 5–10 October 1997. N.A. Hanan, S.C. Mo, and J.E. Matos, "Transient Analysis for HEU and LEU Designs of the FRM-II," paper presented at the 1998 International Meeting on Reduced Enrichment for Research and Test Reactors, São Paulo, Brazil, 18–23 October 1998.

10 For example, see Mo, et al., "Comparison of the FRM-II." Hanan, et al., "An Alternative LEU Design." Alexander Glaser, "Neutronics Calculations Relevant to the Conversion of Research Reactors to Low-Enriched Fuel," Ph.D. diss., Technischen Universität Darmstadt, 2005.

11 The RERTR group at ANL made contributions that went into developing the conversion strategies to consider.

12 MEU is colloquially defined as uranium enriched to at least 20 percent but no more than 50 percent. But MEU is not a legally recognized designation, and uranium of between 20–50 percent enrichment is still strictly HEU.

13 "Vereinbarung zwischen der Bundesrepublik Deutschland vertreten durch die Bundesministerin für Bildung un Forschung (Bund) und dem Freistaat Bayern, vertreten durch den Bayerischen Staatsminister für Wissenschaft, Forschung, und Kunst (Land)," 2001. "Teilgenehmigung nach §7 Atomgesetz (AtG) zum Betrieb der Hochflussneutronenquelle München in Garching (FRM-II): 3. Teilgenehmigung," Bayerisches Staatsministerium für Landesentwicklung und Umweltfragen, 2 May 2003.

14 A. Röhrmoser, I. Neuhas, C. Müller, P. Jüttner, H. Gerstenberg, and W. Fries, "Neutronics Evaluation of Target Irradiation at FRM-II for Mo-99 Supply," Paper presented at the 12th International Meeting of the International Group on Research Reactors, Beijing, China, 27–31 October 2009.

15 "Neues Fingerhutrohr für Molybdän-99-Produktion erfolgreich eingebaut," 18 February 2011, http://www.frm2.tum.de/de/aktuelles/news/einzelnews/article/neues-fingerhutrohr-fuer-molybdaen-99-produktion-erfolgreich-eingebaut/back/1661/index.html (accessed on 27 September 2012).

16 Christian C. Müller, "Radioisotopes for Europe," *FRM-II News*, No. 5, December, 2010.

17 Christiane Funke, "Reaktor läuft länger mit waffenfähigem Uran," *Sueddeutsche Zeitung*, 26 July 2010.

18 Alan J. Kuperman, personal communication, 5 July 2012. It also is unclear specifically how the 2010 deadline for conversion to MEU has been legally bypassed.

19 K. Böning, Winfried Petry, A. Röhrmoser, Chr. Morkel, and N. Wieschalla, "Conversion of the FRM-II," paper presented at the 8th International Topical Meeting, Research Reactor Fuel Management, Munich, Germany, 21–24 March 2004, states that the "risk of proliferation of HEU fuel elements [is] extremely small as compared to other scenarios of proliferation." This suggests that the operators of the FRM-II do not share global concerns about the security risks associated with HEU fuel, as enunciated at the first two nuclear security summits in 2010 and 2012.

20 Glaser, "Neutronics Calculations."

21 M. Englert, "Neutronenphysikalisch Simulationsrechnungen zur Proliferationsresistenz nuklearer Technologien" (PhD diss., Technischen Universität Darmstadt, 2009).

22 The density of uranium metal is 19.1 g/cc, thus imposing an upper bound on the density of fuel meats achievable.

23 Patrick Lemoine and Dan Wachs, "High Density Fuel Development for Research Reactors," preprint of a paper presented at GLOBAL 2007, also refers to work on UMo dispersion and monolithic fuels by "CNEA in Argentina, AECL in Canada, CEA in France, TUM in Germany, KAERI in Korea, VNIIM, RDIPE, IPPE, NCCP, and RIARR in Russia, and INL, ANL, and Y-12 [in the United States]." Further references regarding the specifics of monolithic UMo fuel fabrication and the results of irradiation tests may be found in their citations.

24 Email correspondence with Winfried Petry.

25 Englert, "Neutronenphysikalisch Simulationsrechnungen."

26 The magnitude and nature of these losses is known from the computational simulations, but the significance of these losses on the experimental utility of the reactor as a constant neutron source and a scientific investment is open for debate.

27 "Teilgenehmigung nach." "Vereinbarung zwischen."

28 Interview with Annette Schaper by author, Frankfurt, Germany, 24 February 2011. It should be noted that this was expressed as her impression and is *not* an expert legal opinion.

29 Interview with Wolfgang Liebert, Darmstadt, Germany, 24 February 2011.

30 Interviews with Winfried Petry, 22 February 2011 and 7 May 2011. Interview with Stefan Kern, Bonn, Germany, 28 February 2011.
31 Interviews with Winfried Petry.
32 Interview with Wolfgang Liebert.
33 See, for example, Glaser, "Neutronics Calculations," or Hanan, et al., "An Alternative LEU Design."
34 Similarly, the increase in the decay heat of the spent fuel after 500 years of storage is of little concern when put into the context of all nuclear waste. This was pointed out in the interview with Annette Schaper.
35 See International Atomic Energy Agency. *IAEA Safeguards Glossary*, IAEA/SG/INF/1, 1980, p. 22.
36 Interview with Annette Schaper.
37 Interview with Wolfgang Liebert.
38 Interview with Matthias Englert, via Skype, 6 December 2010.
39 Comprehensive National Energy Policy Act of 1992. H.R. 776. 102nd Cong., 2nd session.
40 Interview with Annette Schaper.
41 Alan J. Kuperman, personal communication, 6 July 2012.
42 Interview with Winfried Petry.
43 One proposed strategy for the final disposal of HEU in some states is to melt and dilute the spent fuel. Such a project is enormously expensive and complex, and would benefit from international scientific and economic cooperation. Interview with Alexander Glaser, via telephone, 18 February 2011.
44 Interview with Alexander Glaser.
45 Interview with Annette Schaper.

# 9 Canada and Russia

## Medical Isotope Production

*Kendra Vessels*

Molybdenum-99 (Mo-99) is the isotope used for the production of technetium-99m (Tc-99m), a medical isotope necessary for about 18 million diagnostic nuclear medicine studies annually in the United States.[1] Tc-99m is the most widely used radioactive isotope for medical diagnostic studies, such as evaluating the medical condition of the heart, kidneys, lungs, liver, and other vital organs.[2] Mo-99 is typically produced in the core of nuclear reactors – using highly enriched uranium (HEU) for the fuel that produces the neutrons, or for the targets that the neutrons irradiate to produce the isotope, or for both.

Such use of HEU is of concern to the security and nonproliferation communities because this fissile material can also be used to make nuclear weapons, by states or terrorists. By contrast, the alternative of low-enriched uranium (LEU) is deemed far safer because the lower concentration of U-235 would necessitate a much larger amount of uranium to make a nuclear weapon. Despite the security risks associated with HEU, however, the world's three largest producers of Mo-99 have yet to convert to using LEU – citing economic, technical, and political reasons.

Traditionally, the world's largest producer of Mo-99 has been Canada's MDS Nordion. Despite Canada's professed dedication to nonproliferation, this Canadian company continues to use HEU targets for production because it deems the costs associated with conversion too high. Moreover, in 2010, MDS Nordion signed a cooperation agreement with the Russian company ISOTOPE – a subsidiary of Russia's State Nuclear Energy Corporation, Rosatom – under which the Russians will produce and supply Mo-99 to Nordion for at least 10 years, even though the Russians use HEU for both fuel and targets. The agreement will enable Russia to expand its isotope production using HEU, just at a time when the rest of the world is phasing out HEU for the production of Mo-99 due to terrorism and proliferation concerns.

This chapter investigates the prospects for converting Russia's Mo-99 production from the use of HEU fuel and targets to LEU alternatives. The question will be addressed by examining the worldwide Mo-99 market, planned expansion of production in Russia, and the potential challenges to converting – whether technical, economic, or political. The chapter concludes with policy recommendations for the United States, Canada, and Russia to facilitate the phase-out of HEU for the production of Mo-99.

## Background

Mo-99 is traditionally produced using a target, a nuclear reactor to irradiate the target, and a processing facility to separate the Mo-99 from the irradiated target (see Figure 9.1). Today, most isotope-production reactors use LEU fuel to produce neutrons, although there are exceptions. But the three largest isotope producers all use HEU *targets*. The last of the world's four major producers, South Africa, is bucking this trend by converting to LEU targets, but it is still in the early stages of large-scale production without HEU. (For details on South Africa's progress, see this book's chapter by Chloe Colby.)

Currently, 99 percent of Russia's demand for Mo-99 is met by production at Obninsk, just outside of Moscow, using a reactor and a processing facility at the Karpov Institute of Physical Chemistry. This facility has been in use since 1985. The Karpov WWR-TS research reactor uses HEU fuel and irradiates HEU targets fabricated at the Institute.[3] The resulting Mo-99 is intended for domestic use only and therefore Russia has been considered a "regional producer," rather than a "major producer."[4]

In recent decades, Canada's MDS Nordion supplied the largest share of the world's Mo-99, including about 60 percent of the U.S. market.[5] At times, it provided the entire U.S. supply of Mo-99. MDS Nordion is a private company, but its raw Mo-99 is actually produced by a government-owned, crown corporation,[6] Atomic Energy of Canada Limited (AECL). AECL fabricates targets using HEU supplied by the United States, irradiates them in the LEU-fueled National Research Universal (NRU) reactor, and extracts Mo-99 from them in a processing plant located near the reactor in Chalk River, Ontario. MDS Nordion then purifies and prepares the Mo-99 before distributing it to the United States and Europe.

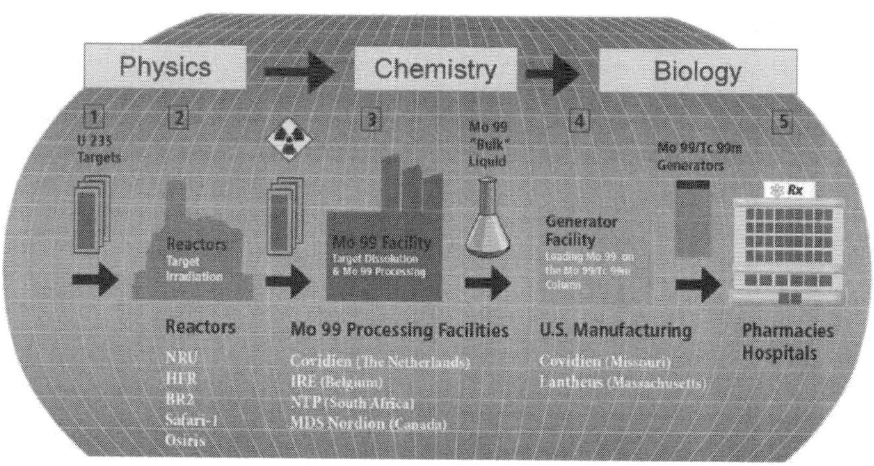


*Figure 9.1* Producing Medical Isotopes: From Targets to Pharmaceuticals

Source: Courtesy TRIUMF, based on a graphic by Nordion, Inc.

This process is straightforward, but it has experienced several setbacks over the past few years due to the age of the NRU and related safety concerns. The NRU began operating in 1957,[7] and AECL has managed to have it relicensed by the Canadian Nuclear Safety Commission (CNSC) every five years despite the reactor's age. However, the NRU was shut down in May 2009 after the discovery of a small leak of heavy water, and it did not return to operation until August 2010, creating a worldwide shortage of medical isotopes. In January 2011, Nordion's vice-president and chief nuclear officer Hank Drumhiller told the CNSC that he expects isotope production at NRU to end in 2016. "That is the decision the government has made."[8]

Because the NRU has been unreliable for some time due to shutdowns, MDS Nordion has sought alternative methods of obtaining Mo-99 for its customers. Among these alternatives is the MAPLE project. Under contract for MDS Nordion, AECL constructed two new MAPLE (Multipurpose Applied Physics Lattice Experiment) reactors and a new processing facility, originally expected to start up in 2000 to replace the NRU and its associated processing facility for production of Mo-99. The MAPLE project – which was controversially designed to use HEU targets despite AECL's earlier pledge to convert to LEU targets by 2000 – would have had the capacity to meet worldwide demand of Mo-99 if needed.[9] However, the Canadian government canceled the project in 2008, after completion of the reactors, due to concerns about safety and performance.[10] MDS Nordion continued to fight for the MAPLE project to be resuscitated,[11] but in September 2012 lost its arbitration effort against AECL.[12] Experts are highly skeptical that the MAPLE project will be revived.[13]

Canada is now pursuing alternative Mo-99 production technologies, which may prove sufficient to meet domestic demand for medical isotopes.[14] But Canada has no successor reactor, nor other technology, capable of large-scale Mo-99 production for export after 2016. Moreover, neither Mo-99 nor Tc-99m can be stockpiled due to their short half-lives: 66 hours and 6 hours, respectively. Therefore, MDS Nordion must find an alternative source of Mo-99 within the next few years to remain a major supplier in the global market.

## Nordion-ISOTOPE 2010 Agreement

In response to the prolonged shutdown of the NRU in 2009–2010, and the uncertainty with the MAPLE project, Canada's Nordion in September 2010 entered into a 10-year agreement with Russia's ISOTOPE.[15] ISOTOPE will supply Mo-99 to MDS Nordion on an exclusive basis for processing, distribution, and sale outside of Russia. In its 2010 annual report, Nordion explained: "While we continue to seek a reliable long-term supply of medical isotopes, this agreement strengthens Nordion's back-up isotope supply and partially offsets the impact of AECL's planned NRU reactor shutdowns."[16] Figure 9.2 depicts Russia's organization of Mo-99 production.

In order for ISOTOPE to fulfill the agreement, Russia is expanding production using facilities at the Research Institute of Atomic Reactors (RIAR or NIIAR),[17]

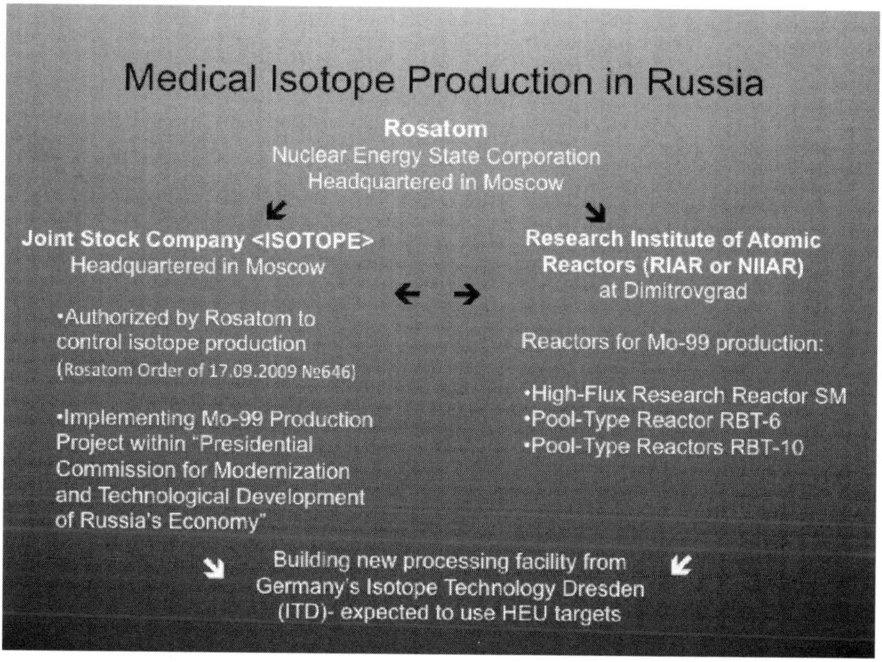


*Figure 9.2* Russian Institutions Involved in Mo-99 Production

in Dmitrovgrad, 500 miles east of Moscow. ISOTOPE plans to use three reactors to irradiate targets that will then be fed into one processing plant, which will extract Mo-99 for export to North America, where Nordion will distribute it to medical providers for use of its decay product, Tc-99m.

RIAR plans to use HEU targets throughout a three-phase expansion of production at Dmitrovgrad, as summarized in Figure 9.3. In the first phase, reportedly completed by January 2011, the RBT-6 reactor and an existing processing facility were to start producing up to 1,000 six-day Ci per week, while construction proceeded on a new, larger ITD processing facility.[18] During 2011, in phase two, ISOTOPE planned to expand production gradually to about 2,700 Ci weekly – roughly 20 percent of global demand – using the RBT-10 reactor, a newly designed target, and the new ITD processing facility with a proprietary "dry process." However, that schedule has slipped by a year or more, as only one of the two planned production lines opened in 2011, while the second was delayed until June 2012.[19] In phase three, originally projected to start in 2012 but now reportedly delayed, Russia plans regularly to produce at least 20 percent of world demand for Mo-99, relying on as many as three reactors: the SM, RBT-6, and RBT-10. According to an August 2012 report, RIAR now expects to produce 800 Ci weekly by 2013, expanding to 2,000 Ci by 2015.[20]

**Project Phases**

- Phase 1: Start-up (by Jan 2011)
  - Irradiation in RBT-6 Reactor
  - TPE rod-type target
  - Produce up to 1,000 Ci/week
  - Construct ITD processing facility
- Phase 2: Increase Production (during 2011)
  - Irradiation in RBT-10; RBT-6 is backup
  - Use new ITD processing facility, new target, and proprietary "dry process"
  - Gradually increase output to 2,700 Ci/week
- Phase 3: Global Supplier Basis (starting 2012)
  - Production capacity of 2,700 Ci/week
  - Capture 20+ % of world market
  - Explore conversion of target from HEU to LEU

*Figure 9.3* ISOTOPE's Original Schedule for 3-Phase Expansion

## Prospects for Conversion

The prospects for converting Canada's targets from HEU to LEU are not assessed in this chapter. As noted, AECL's NRU reactor is scheduled to stop producing Mo-99 in 2016. There are no plans, or demands, to convert its targets before then, in light of the short timeline. The MAPLE project is canceled and there are no other reactors in Canada slated for large-scale Mo-99 production. Moreover, Canada's government has indicated that any new production of Mo-99 in the country will avoid HEU.[21] Accordingly, it appears that Canada will cease using HEU for production of Mo-99 by 2016, at the latest, despite its past refusal to convert its targets to LEU.

To assess the prospects of Russia converting to LEU targets, it is necessary to explore the technical, economic, and political challenges. Such an assessment ideally would rely on primary sources including interviews with officials in Russia, especially at ISOTOPE. However, such primary sources were inaccessible. Accordingly, this chapter relies mainly on secondary sources, including official statements and reports from ISOTOPE and RIAR.

Regarding potential conversion from HEU to LEU, RIAR's Deputy Director Rostislav A. Kuznetsov told a conference, in August 2010, that it is possible in Phase 3 of expansion.[22] Likewise, in a March 2011 meeting at the Embassy of Russia, in Washington, Rosatom's Director General Sergei Kiriyenko reportedly said that conversion to LEU targets is feasible, but he refused to make a commitment – or propose a timeline – for conversion.[23] Most encouraging, Russian officials report that on 8 July 2012, Kiriyenko "approved a plan for Russia to convert to LEU targets for Mo-99 production instead of HEU targets within 5 years."[24] In October 2012, Rosatom's deputy director announced the creation of such a conversion program.[25] The scientific literature concurs that there are no technical obstacles to Russia converting to LEU targets.[26]

Until recently, there may have been Russian political obstacles to conversion, but those appear to have faded. For example, until 2010, Russia resisted even exploring conversion of the fuel in its research reactors from HEU to LEU. But since then, Russia has engaged in feasibility studies on converting six of its reactors, and appears ready to implement its first conversion in 2014. (For details, see this book's chapter by Braden Civins.) Moreover, Russia has embraced multilateral statements favoring HEU minimization at the first two Nuclear Security Summits, in 2010 and 2012, in Washington and Seoul.

Rather than technical or political obstacles, it appears that economic concerns drive Russia's reluctance to convert to LEU targets. ISOTOPE reportedly wants to expand its market share quickly, instead of delaying such expansion while it converts to LEU targets, which could take several years. After it carves out a share of the market, the company could prove more willing to consider conversion to LEU, but only if there were some incentive to do so, as discussed below.

## Policy Recommendations

### *Canada*

The Canadian government owns AECL and therefore had leverage in the past to encourage Nordion to convert to LEU targets, but it never did so due to concerns about a potential shortage of Mo-99 during the conversion process.[27] Given that conversion to LEU targets now is unlikely at the NRU before the reactor stops producing Mo-99 in 2016, Canada's government should instead focus on working with Rosatom to convert ISOTOPE's production process from HEU to LEU. Considering that Nordion is to purchase all of the Mo-99 produced at RIAR for export, Canada should use its influence with Nordion to convince ISOTOPE to convert to LEU – at least for the targets, if not also the reactor fuel, used to produce Mo-99.

### *United States*

In the United States, the January 2013 enactment of the American Medical Isotope Production Act could make a difference in both Russia and Canada. This legislation would promote the production of Mo-99 without HEU in the United States, and phase out the export of HEU for the production of Mo-99 elsewhere. As the U.S. National Nuclear Security Administration (NNSA) official Parrish Staples explained to the Senate Energy and Natural Resources Committee in February 2011:

> Provisions of this legislation … are aligned with the NNSA's nonproliferation mission to assist in the conversion of research reactors and isotope production facilities worldwide from the use of HEU to LEU, and to establish a reliable supply of Mo-99 produced without the use of HEU in the United States.[28]

MDS Nordion, in its 2010 annual report, recognized the potential of this legislation to significantly impact its market in the United States.[29]

Even prior to enactment of the legislation, the NNSA's Global Threat Reduction Initiative was already making efforts to minimize the use of HEU for medical isotope production.[30] In response to a question from the Senate committee about whether the United States has engaged Russia concerning HEU use for Mo-99 production, Parrish Staples stated in 2011 that "We intend to raise this issue at very high levels, in meetings with various Rosatom officials such as Director Kiriyenko, to dissuade Russia from the use of HEU to ensure that their actions are consistent with the presidential-level commitments that they've made at the April 2010 nuclear security summit."[31]

In the boldest initiative so far to phase out HEU for production of Mo-99, the United States announced in July 2012 that it intends to pay more for Mo-99 produced without HEU.[32] This plan would cover only medical isotopes reimbursed with federal funds, such as Medicare, and by itself might not persuade Russia to convert. But the White House also announced its intention "to promote the conversion of all medical isotope production to non-HEU sources," and reportedly is considering further steps to achieve that goal.[33]

### *Russia*

Russia appears committed to agreements to reduce civilian use of HEU in areas other than medical isotope production.[34] Expanding that cooperation to also convert Mo-99 production to LEU targets could potentially have economic benefits for Russia. Although the recently enacted American Medical Isotopes Production Act does not directly restrict the U.S. import of medical isotopes produced with HEU, the prospect of such additional provisions has been cited repeatedly.[35] Even prior to enactment of further legislation, the White House initiative to pay more for Mo-99 produced without HEU might well discourage imports of Russian Mo-99 produced with HEU. In the face of such challenges, Russia can ensure access to the U.S. market only by converting its Mo-99 production process to use LEU targets as soon as possible.[36] Even if ISOTOPE insists on postponing conversion until Phase 3 of its ongoing expansion of production, to first capture market share, this chapter recommends that Russia make a commitment now to eventual LEU conversion, to help ensure access to markets that may have a preference for Mo-99 produced without HEU, such as the United States.

## Conclusion

This chapter has focused on the prospects of Russia converting from HEU to LEU targets for production of Mo-99, in the wake of the 2010 agreement under which Russia will produce Mo-99 for worldwide distribution by the Canadian company Nordion. By reviewing the current production processes, expected expansion in Russia, and feasibility of conversion, the chapter demonstrates that Russia has considered conversion, determined that it is feasible, but refused so far to commit to it. If the United States and Canada engage Russia to commit now to convert, at least during the third phase of Russia's expansion of production, Nordion and ISOTOPE could help ensure their future access to the U.S. market, even as Washington adopts policies that penalize and restrict the use of Mo-99 produced with HEU.

## Notes

1 J. Brice, "New Mo-99 shortage threatens following Chalk River reactor shutdown," *Diagnostic Imaging*, 20 May 2009, http://www.diagnosticimaging.com/nuclear/content/article/113619/1415403 (accessed on 10 April 2011).

2 "Technetium-99," U.S. Environmental Protection Agency, 24 March 2011, http://www.epa.gov/rpdweb00/radionuclides/technetium.html (accessed on 27 September 2012).

3 National Research Council, *Medical Isotope Production Without Highly Enriched Uranium* (National Academies Press, 2009), 34. This reactor is sometimes called the "VVR-T."

4 Regional producers manufacture Mo-99 primarily for domestic or regional use and each supply considerably fewer than 1,000 6-day curies per week, collectively producing only about 5 percent of the world supply. Major producers manufacture more than 1000 6-day curies of Mo-99 per week. National Research Council, *Medical Isotope Production*, 33–34.

5 National Research Council. *Medical Isotope Production*, 34.

6 A Crown Corporation is owned by the executive branch of Canada's government.

7 "The NRU Reactor," National Research Council Canada, http://www.nrureactor.ca/index.html (accessed on 27 September 2012).

8 T. Meyers, "NRU helps put Nordion back in black," *North Renfrew Times.com*, 9 February 2011, http://www.magma.ca/~drcanrt/110209nordion.html (accessed on 10 April 2011). The reactor may remain in use longer for other purposes such as research and to support AECL's commercial reactor program.

9 Alan J. Kuperman, "Bomb-grade Bazaar," *Bulletin of the Atomic Scientists*. 62,2 (2006): 44–50.

10 "AECL aborts development work on MAPLE medical isotope reactors," *Canadian Press*, 16 May 2008. Alison Motluk, "Nuclear safety: When positive is negative," *New Scientist*, 19 January 2010.

11 "MDS urges Canada to restart Maple reactor project," *Reuters*, 11 June 2009.

12 Bert Hill, "Nordion needs win in Chalk River reactor arbitration," *Ottawa Citizen*, 19 June 2012. Tim Lougheed, "One Last Nail in the MAPLEs Coffin," *Canadian Medical Association Journal*, 12 September 2012.

13 Jean-Pierre Labrie, "Don't count on MAPLE to deliver medical isotopes," *National Post*, 28 July 2009.

14 Mark Hume, "In a breakthrough, Canadian researchers develop a new way to produce medical isotopes," *Globe and Mail*, 20 February 2012. Tim Lougheed, "Uncertainties shroud medical isotope supply," *Canadian Medical Association Journal*, 14 June 2012.

15 "Final Transcript: MDZ–MDS Inc. to Discuss Signing of Framework Agreement with JSC Isotope – Conference Call," 27 September 2010, http://www.nordion.com/reports/MDZ-Transcript-2010-09-27T15_00.pdf (accessed on 27 September 2012).

16 Nordion Inc., "Fiscal 2010 Annual Report," 3. The report adds: "The restart of the MAPLE reactors, however, remains our preferred long-term supply solution."

17 RIAR is also known as NIIAR, the acronym of the organization's name in Russian.

18 Production at RIAR started on 18 December 2010, according to Dmitry Konukhov, Center for Energy and Security Studies, "Report On the 7th International Conference

on Isotopes, 4–8 September 2011, Moscow, Russia," but this appears to indicate test production.

19 Konukhov, "Report On the 7th International Conference on Isotopes," reports that one new production line started in summer 2011, and the second was scheduled to start in 2012, to achieve full capacity. In June 2012, ISOTOPE announced that the second line was to start by the end of the month. See "Пусконаладка второй очереди производства Mo-99 в НИИАР начнется в конце июня," 5 June 2012, http://www.isotop.ru/view/667/ (accessed on 1 August 2012).

20 OECD Nuclear Energy Agency (NEA), "A Supply and Demand Update of the Molybdenum-99 Market," August 2012, p. 10, www.oecd-nea.org/med-radio/docs/2012-supply-demand.pdf (accessed on 27 September 2012).

21 Nordion Inc., "Fiscal 2010 Annual Report," December 2010, 4, http://www.nordion.com/reports/2010_engannual.pdf (accessed on 27 September 2012).

22 "Consultancy on Conversion Planning for Mo-99 Production Facilities from HEU to LEU," August 2010, http://www.iaea.org/OurWork/ST/NE/NEFW/Technical_Areas/RRS/documents/mo99/WORKINGMATERIALSmo99CM.pdf (accessed on 27 September 2012).

23 Miles Pomper, email to Alan J. Kuperman, 22 April 2011.

24 Miles Pomper, email to Alan J. Kuperman, 9 August 2012, citing minutes of the July 2012 meeting of the High-level Group on the Security of Supply of Medical Radioisotopes (HLG-MR) of the OECD Nuclear Energy Agency.

25 Pavel Podvig, "Russia to begin conversion of its research reactors to LEU in 2013," http://fissilematerials.org/blog/2012/10/russia_to_begin_conversio.html (accessed on 13 November 2012).

26 National Research Council. *Medical Isotope Production*, 34.

27 Kuperman, "Bomb-grade Bazaar."

28 Parrish Staples, "Testimony before the Senate Committee on Energy & Natural Resources," 1 February 2011, http://www.nnsa.energy.gov/mediaroom/congressional testimony/staplestestimony2111 (accessed on 27 September 2012).

29 Nordion Inc., "Fiscal 2010 Annual Report."

30 Parrish Staples, "The NNSA Global Threat Reduction Initiative's Efforts to Minimize the Use of Highly Enriched Uranium for Medical Isotope Production," RRFM 2010, Marrakech, Morocco, 21–25 March 2010, http://www.iaea.org/inis/collection/NCLCollectionStore/_Public/41/064/41064165.pdf (accessed on 27 September 2012).

31 U.S. Senate Committee on Energy And Natural Resources, Hearing On S. 99, the American Medical Isotopes Production Act of 2011, Senate Hearing 112–4, 1 February 2011, http://www.gpo.gov/fdsys/pkg/CHRG-112shrg65421/html/CHRG-112shrg65421.htm (accessed on 27 September 2012).

32 Douglas P. Guarino, "U.S. Proposes Bigger Medicare Payouts to Avoid Bomb-Grade Uranium," Global Security Newswire, 11 July 2012, http://www.nti.org/gsn/article/us-proposes-bigger-medicare-payouts-avoid-bomb-grade-uraniu/ (accessed on 27 September 2012).

33 Douglas P. Guarino, "Obama Administration Weighing More Aggressive Medical Isotope Policies," Global Security Newswire, 11 June 2012, http://www.nti.org/gsn/article/obama-administration-weighing-more-aggressive-medical-isotope-policies/ (accessed on 27 September 2012).

34 Douglas P. Guarino, "NNSA Considers Clarifying Statement on Russian Isotope Commitment," Global Security Newswire, 26 June 2012, http://www.nti.org/gsn/article/nnsa-considers-clarifying-statement-russian-isotope-commitment/ (accessed on 27 September 2012).

35 In late 2010, the CEO of South Africa's Mo-99 producer was quoted as saying that he expected the United States to erect "some kind of tariff barrier which would make it not as attractive to import molybdenum[-99] … derived from highly-enriched

uranium," Benjamin Humphrey, "S. Africa turns apartheid-era nukes into medicine," *Agence France Presse*, 12 November 2010. See also, Guarino, "Obama Administration Weighing More Aggressive Medical Isotope Policies."

36 Miles A. Pomper and William C. Potter, "Medical isotope production: The US must follow South Africa's lead," *Bulletin of the Atomic Scientists*, 17 December 2010, http://www.thebulletin.org/web-edition/features/medical-isotope-production-the-us-must-follow-south-africas-lead (accessed on 27 September 2012).

# 10 Russia

## Research Reactors

*Braden Civins*

Russia has long been the world's leading user of HEU to fuel research reactors.[1] Partially as a result, Russia possesses the world's most civilian HEU, holding between 15 and 30 metric tons, "accounting for more than half of the world's civilian HEU."[2] This chapter addresses the technical, economic, and political prospects and challenges of converting Russia's HEU-fueled, steady-state, research reactors to LEU fuel. (That excludes Russia's critical assemblies and pulsed reactors, which are covered separately in this book's chapter by Paul Osborne.) Eliminating the use of HEU in Russian research reactors would have an immediate, tangible impact on international security and, as others have noted, "help establish an international norm to convince other countries to give due recognition to the threat."[3]

This chapter presents the results of research since September 2010, offering new findings to complement published information on Russia's civilian research reactors. Significant developments in Russian nuclear policy occurred during the course of this research, so this study is one of the first to report these developments and situate them in historical context. The chapter includes the following: an assessment of the Russian State Atomic Energy Corporation's (Rosatom) authority in decision-making processes regarding conversion; perspectives offered by U.S. and Russian scientists and policymakers; and insight into underlying reasons for the Russian government's apparent, recent policy change in favor of converting its own domestic research reactors to LEU fuel.

Part I of this chapter surveys available information regarding Russian HEU-fueled research reactors and provides an overview of US-Russian bilateral cooperation on nuclear-security issues since the early 1990s. Part II provides the methodology used to research technical, economic, and political obstacles to conversion. Part III reports the findings of this research. Part IV lays out important lessons learned. Part V offers conclusions, makes policy recommendations for expediting the conversion process in Russia, and suggests additional topics meriting further research.

### Literature Review

In the 1950s, the Soviet Union and United States exported civilian research reactors to their respective allies. Both countries also exported HEU to fuel these research reactors in the 1960s and 1970s. In response to demand by reactor operators in Soviet bloc states for more powerful fuel, the Soviets exported HEU enriched to 90 percent.[4] Both the United States and Soviet Union became increasingly concerned about global proliferation of weapons-grade fissile material in the late 1970s, and both countries implemented externally focused programs to reduce HEU exports. The U.S. program, RERTR, and the Soviet program, RRFR, were both designed to reduce the enrichment of fuel in civilian research reactors located abroad. The Soviet program, RRFR, sought conversion of HEU reactors to use of uranium enriched to 36 percent – which by virtue of being enriched beyond 20 percent, is still considered HEU.

In 1993, the U.S. and Russia initiated "the first formal [bilateral] contact to discuss collaboration on research reactor conversions."[5] As a result of this meeting, the U.S. Argonne National Laboratory, and Russia's Research and Development Institute of Power Engineering (NIKIET), signed a contract as a basis for collaboration on conversion studies and fuel development projects.[6]

Since bilateral cooperation on civilian HEU minimization began, the US and Russia have partnered on a host of nonproliferation-oriented initiatives, including those aimed at securing and reducing quantities of HEU in Russia and elsewhere,[7] and combating nuclear smuggling and terrorism.[8] Since the establishment of DOE's Global Threat Reduction Initiative (GTRI) in 2004, DOE and Rosatom have partnered to convert remaining Russian-supplied research reactors located outside Russia. GTRI continues to support Russian efforts to develop and qualify "LEU-Mo fuel designs that will enable conversion of at least 8 HPRRs [high-performance research reactors] in Russia."[9]

In 1986, the U.S. NRC ordered conversion of all licensed domestic research and test reactors, and eventually DOE began exploring the possibility of converting its own reactors. As of 2010, "17 university and 2 DOE research reactors in the U.S. [already had] been converted,"[10] and by 2012 another reactor was too, leaving only about seven HEU-fueled U.S. research reactors.[11] Russia, on the other hand, had until very recently not pursued a domestic conversion policy.[12]

Despite the increasing sophistication and worldwide use of high-density LEU fuel in research reactors, Russia had not implemented a coherent conversion policy for several known reasons, such as:

> the apparent high short-term financial costs of converting to LEU, perceived advantages offered by the use of HEU for some advanced research projects, fear of losing a potential technological edge vis-à-vis other countries … as well as other economic, social, and political factors. An additional challenge in gaining Russia's support for a civilian HEU phase-out is the Russian perception that the threat of terrorists acquiring HEU for the purpose of fabricating an improvised nuclear device is not very high.[13]

The phase-out of HEU in Russia's research reactors would carry an economic cost to some domestic actors regardless of the reactors' ultimate disposition – i.e., converted or shut down. Rosatom has stated that simply shutting down reactors would incur costs arising from the decommissioning and disposal of fresh and spent fuel.[14] Social factors also stand in the way of conversion, as HEU reactors "may be considered a source of prestige in Russia with institutions reluctant to close the door to future nuclear research that may require HEU."[15]

Several technical barriers remain despite the increasing feasibility of conversion. Rosatom has suggested previously that the development of suitable LEU-based fuels is a primary challenge,[16] particularly as research reactors installed in various facilities are diverse in design, power levels, fuel composition, and operation mode.[17] But an independent expert, Elena Sokova, downplays technical hindrances to conversion, arguing that "Technological barriers to conversion should be all but removed by 2011, when new high-density fuels become available and a number of Russian reactors are near the end of their service lives."[18]

Estimating the number of operational research reactors in Russia is complicated "by the fact that the reactors … are owned by various Russian entities, including academic institutions and the ministries of Atomic Energy, Education, Health, and Defense."[19] In June 2012, DOE reported that "nine out of 27 research reactors using highly enriched fuel in the Russian Federation have been shut down," implying that 18 such reactors were still operating.[20] Less than two years earlier, RERTR had estimated there were 16 HEU research reactors in Russia that used a variety of fuel geometries, including tubular, rods, and liquid (solution).[21] Of these, "11 civilian reactors using aluminum-based fuels in tubes or pins are the type of reactors for which LEU fuels have been developed or are under development in both the US and Russian RERTR programs," and the one liquid-core research reactor is also a candidate for LEU conversion.[22] At least four high-performance research reactors may not be convertible to LEU fuels that are currently available or under design.[23] A single fast research reactor, although it cannot be converted, is scheduled to shut down in 2015.[24] A list of reactors is provided in Table 10.1.

In light of Russia's traditional reluctance to convert domestic research reactors, several recent developments are noteworthy. In 2010, after protracted negotiations with the United States, the Russian government agreed to conduct feasibility studies for the possible conversion of six HEU-fueled research reactors.[25] The terms of the feasibility studies were finalized in an Implementing Agreement, signed in December 2010 by Rosatom's Director General, Sergey Kiriyenko, and U.S. Deputy Secretary of Energy, Daniel Poneman.[26] They are being conducted as a collaborative effort by the U.S. Argonne National Laboratory, Rosatom, and various Russian nuclear facilities.

In addition, at the 32nd International Meeting on Reduced Enrichment for Research and Test Reactors (RERTR) in Lisbon, in October 2010, Nikolai Arkhangelskiy, a senior Rosatom official, announced the closure of three "critical assemblies," which are zero-power facilities that typically contain HEU and are used to design reactors.[27] The significance of Arkhangelskiy's announcement is unclear, as one of the indicated facilities had been categorized as inoperative some time ago.[28] He also hinted at the scheduled decommissioning of several other, unnamed facilities. (For more on Russian critical assemblies, see the chapter in this volume by Paul Osborne.)

*Table 10.1* Status of Steady-State, HEU-Fueled, Research Reactors in Russia (excluding those already shut down permanently)

| *Reactor* | *Operator* | *Location* | *Enr %* | *MW* | *1st Crit.* | *Notes* |
|---|---|---|---|---|---|---|
| *Conversion Feasibility Studies Completed (or underway)* | | | | | | |
| MIR-M1 | RIAR | Dmitrovgrad | 90 | 100 | 1966 | |
| IR-8 | RRC Kurchatov Institute | Moscow | 90 | 8 | 1981 | |
| IRT-T | Tomsk Polytechnic University | Tomsk | 90 | 6 | 1967 | |
| IRT | Moscow Engineering Physics Institute | Moscow | 90 | 2.5 | 1967 | Shut for upgrades |
| ARGUS | RRC Kurchatov Institute | Moscow | 90 | 0.02 | 1981 | Liquid core |
| OR | RRC Kurchatov Institute | Moscow | 36 | 0.3 | 1988 | |
| *Conversion Feasibility Likely but Needs Study* | | | | | | |
| WWR-M | Petersburg Nuclear Physics Institute | St. Petersburg | 90 | 18 | 1959 | aka, VVR-M |
| IVV-2M | Institute of Nuclear Materials | Zarechniy (Sverdlovsk) | 90 | 15 | 1966 | |
| WWR-TS | Karpov Institute of Physical Chemistry | Obninsk | 36 | 15 | 1964 | |
| GAMMA | RRC Kurchatov Institute | Moscow | 36 | 0.1 | 1981 | Not shut, as reported |
| IRV-2M | Research Inst. Of Scientific Instruments | Lytkarino (Moscow) | 90 | 4 | ~2016 | Re-construction of old IRV-1M |
| *Conversion Feasibility Uncertain* | | | | | | |
| PIK | Petersburg Nuclear Physics Institute | St. Petersburg | 90 | 100 | 2011 | |
| SM-3 | RIAR | Dmitrovgrad | 90 | 100 | 1961 | aka, CM-3 |
| RBT-10/2 | RIAR | Dmitrovgrad | 90 | 10 | 1984 | |
| RBT-6 | RIAR | Dmitrovgrad | 90 | 6 | 1975 | |
| *Conversion Feasibility Unlikely* | | | | | | |
| BOR-60 | RIAR | Dmitrovgrad | < 90 | 60 | 1969 | Fast reactor w/ Pu also in core |

Four of the conversion feasibility studies were completed by June 2012 – for the Argus, IRT (at MEPhI), OR, and IR-8 – and according to DOE the studies "confirm that it is technically possible to convert the reactors to use low-enriched uranium (LEU) fuel."[29] The two remaining feasibility studies were expected to be completed in summer 2012. Based on the initial studies, Russia has informed the United States that "one or two reactors will be converted in 2014," according to a Russian press report.[30] The IRT and the Argus liquid-core reactor are the two that may convert in 2014.[31] The combined cost of the six conversions is estimated at no more than $12.5 million, excluding the cost of the new LEU fuel, according to Kiriyenko.[32]

These developments seem to demonstrate an attitudinal shift by the Russian government in favor of eliminating the use of HEU fuel in research reactors. Such a shift would be particularly significant in light of past reports that "Representatives of GTRI say they have been approached by numerous Russian facility operators who are interested in conversion assistance but have been told by Rosatom that they must wait for approval from the Russian presidential administration."[33] It remains uncertain as to whether the Russian government has adopted a categorical conversion policy toward its domestic reactors or merely plans to decommission and explore conversion selectively. As of this writing, Rosatom had not formally committed to convert any of its domestic research reactors.[34] But in October 2012, Rosatom's deputy director announced the creation of a program to convert the fuel for research reactors, and the targets for production of medical isotopes, from HEU to LEU.[35]

## Methodology

In addition to secondary sources, this chapter draws upon primary-source interviews with research-reactor operators in Russia, and nuclear engineers and government officials in the United States and Russia, including Nikolai Arkhangelskiy, a senior Rosatom official. Secondary sources and non-published documents provided insight into the technical challenges of conversion. In particular, scientific papers presented at past RERTR conferences highlighted technical challenges in LEU fuel development and suggested specific avenues of inquiry. This chapter's main research questions were as follows:

- Why are reactor operators not converting?
- What are operators' views on conversion?
- What are operators' views on the threats posed by HEU at civilian facilities?
- What incentives would convince operators to convert?
- What is the Russian government's policy towards conversion of its domestic reactors?
- Are there other stakeholders involved in the operation of research reactors who are opposed to conversion?

## Findings

In examining the motivations for Rosatom's newfound interest in pursuing conversion of Russia's domestic research reactors, U.S. experts pointed to emerging global norms. RERTR was credited with driving more favorable international views on HEU minimization. Likewise, a host of diplomatic and legislative efforts – including UN Security Council Resolution 1887 in 2009,[36] the first Nuclear Security Summit in 2010, the resulting four-year multilateral effort to secure all vulnerable nuclear material worldwide, and U.S. congressional initiatives aimed at reducing HEU exports – were cited as helping build norms that may have influenced the Russian government.[37]

However, information gathered during field research in Moscow in January 2011 suggested a more nuanced explanation. An in-person interview with Rosatom official Nikolai Arkhangelskiy offered rare insight into the underlying reasons for the Russian government's pursuit of domestic conversion. Arkhangelskiy cited the persuasive impact of the U.S. government pursuing conversion of high-performance research reactors in the United States and proving that it was feasible. He reported that the Russian government had not even considered the possibility of converting its domestic reactors until the United States showed that it could convert its own. Arkhangelskiy stated:

> To a great extent it was the position of the United States that influenced our position because the U.S. announced that they were going to convert not only low-powered reactors but all reactors and they stated that it was feasible and would not affect in any way the work of researchers … It was proved in practice that penalties [loss of neutron flux] for conversion were minimal.[38]

The U.S. decision to convert domestic, high-performance research reactors also communicated to Russia that HEU minimization was viewed by some international actors as a priority for global security. But it was the U.S. technical progress toward converting high-performance research reactors without causing significant loss of neutron flux that convinced the Russians to undertake domestic conversion.[39]

Arkhangelskiy's explanation for Rosatom's attitudinal shift touched on a recurrent theme throughout the course of this research: the tension in any country between nuclear-security officials – pursuing HEU minimization – and research reactor operators seeking to maximize neutron flux, avoid interruption of operations, and minimize costs. Dr. Dale Klein, former head of the U.S. Nuclear Regulatory Commission, described operators as justifiably resistant to conversion unless a third party, such as the government, can definitively prove that conversion can be accomplished without causing a loss of flux.[40] Operators are largely unconcerned with the proliferation risks associated with HEU and therefore unlikely to be persuaded to convert if there is any risk of loss in flux.

Rosatom's newfound interest in domestic conversion is apparently sincere. Likewise, the December 2010 Implementation Agreement to explore conversion of six research reactors is a significant step toward reducing non-weapons HEU use in Russia. However, feasibility studies are only a first step. As Arkhangelskiy noted, two threshold determinations must be made before conversion can occur: (1) conversion is technically feasible; and (2) conversion is economically justifiable. The Implementing Agreement only addresses the first issue and so does not guarantee conversion. Moreover, a separate agreement would be required to extend bilateral cooperation through the actual conversion process.[41]

Expressing his personal opinion, Arkhangelskiy was optimistic that "in the future there will not be any HEU used in the research reactors." He attributed his optimism to "the evolution going on," referring to the international embrace of HEU minimization. Although normative trends have traditionally played a marginal role in encouraging Russia to phase out HEU, Arkhangelskiy's comments suggest that these trends could play a growing role in sustaining efforts to eliminate Russia's use of HEU for non-weapons purposes.

### *Technical Impediments*

Russia has made progress toward overcoming technical hurdles to conversion, but substantial challenges persist. Development of high-density LEU fuels remains a primary impediment. Arkhangelskiy noted Russian scientists have been developing U-Mo alloy for over a decade and expressed optimism that further development will lead to the conversion of both low-power and high-performance research reactors. However, high-performance research reactors are particularly difficult to convert. U.S. officials note that the potential loss of neutron flux, even if no more than 5 to 10 percent, which is a common estimate under conversion scenarios, can seem like a scientific burden to many operators. As Arkhangelskiy reported, "unfortunately, we have not yet finished work on fuel development … Therefore we cannot say at this point that the issue of conversion of high-powered research reactors is solved."

More specifically, Arkhangelskiy suggested that production of the LEU-Mo dispersion fuel needed to convert certain reactors had been delayed by two to three years due to problems during the qualification process.[42] Nevertheless, he expressed confidence that dispersion fuel, which is the exclusive focus of Russian LEU fuel development efforts, "in principle … should be enough to convert almost all reactors except for the" SM-3 reactor at Dmitrovgrad. Arkhangelskiy believes that the SM-3 also could eventually be converted using monolithic LEU-Mo fuel being developed by the United States.

U.S. RERTR program estimates are less optimistic than Arkhangelskiy's regarding the prospects and timeline of LEU fuel development for Russia's four high-performance research reactors: RBT-10/2, RBT-6, PIK, and SM-3. As the current HEU fuel in these reactors has a high uranium density and enrichment levels of around 90 percent, ultra-high density LEU-Mo fuel will be required for each, and "fuel development, establishing a manufacturing capability, and licensing a new LEU fuel will take 5–10 years or more."[43]

Tests run by the Russian RERTR program illustrate the technical difficulties in the development of U-Mo fuels. Russian scientists have identified specific challenges during development – as have U.S. and French developers – such as curbing the interaction between U-Mo fuel particles and the surrounding aluminum matrix:

> As fuel burns up, the amount of matrix material decreases because of this interaction and significant changes in the fuel meat physical and mechanical properties are observed, including a significant decrease of its thermal conductivity and elevation of its temperature.[44]

The Russian report concludes by suggesting that the addition of silicon to the matrix can limit the interaction. This is only one example of the technical challenges that must be overcome to convert Russian high-performance research reactors to LEU fuel.[45]

### *Economic Impediments*

Interviews and correspondence with Russian nuclear scientists and Nikolai Arkhangelskiy suggest economic concerns drive much of the reluctance to convert. The economic costs associated with conversion stem from different parts of the process. Arkhangelsky specifically cited the high cost of producing high-density LEU fuel, noting that despite hopes that producing it would eventually cost less than HEU fuel, the high-density LEU continues to incur higher production costs. So long as LEU is more costly than HEU, Arkhangelskiy acknowledges, the issue of conversion "cannot be called clear."

When asked to discuss Rosatom's strategy for convincing reactor operators to convert, Arkhangelskiy again emphasized the centrality of fuel cost to the ultimate decision to convert:

> It's too early to speak about negotiations with operators. First we discuss technical feasibility, then we start conversion process … Here, the issue of fuel cost arises. Everything will depend on the cost. LEU fuel will be much more expensive than HEU. At this point, I cannot forecast the decisions.[46]

Of course, Arkhangelskiy acknowledged that low-powered reactors only require a single lifetime core of fuel, easing the economic cost of LEU. Turning to high-performance reactors that require regular supplies of fuel, Arkhangelskiy noted that "even at this point the HEU fuel cost is a burden for operators."[47]

Nuclear reactor operators expressed concerns both about the cost of conversion and who would pay. When asked about various concerns with the conversion process, a senior nuclear researcher at the Karpov Institute of Physical Chemistry at Obninsk repeatedly cited the "economic efficiency" of the conversion process as the ultimate determinant of whether conversion will occur.[48] A Russian reactor operator at Obninsk suggests that it is unclear who ultimately would bear the burden of financing conversion: Rosatom, the facility itself, or some combination thereof.[49]

This same official confirmed previous speculation about another economic obstacle to conversion in Russia – that salaries are directly tied to the type of fuel used at a given facility. He reported:

> Radiation status of the facility is [a] direct influencing factor to the personnel salary. If you just remove nuclear materials from the facility it will cost maybe half of the personnel income. So balance must be always in mind of decision-makers – future of the facility (usefulness for future R&D purpose), cost and time of conversion, efforts and technical capability of the facility to work in new conditions, and a number of other factors.[50]

A U.S. government source highlights other practical hindrances to conversion in Russia. In any country, conversion gives rise to bureaucratic red-tape and costs associated with preparation of regulatory approval documents, safety analysis reports, etc. Russian operators might view the administrative details as a greater deterrent than the actual conversion process, according to this official.[51] Operators may also be concerned that the conversion process could invite unwanted scrutiny of their facilities. As operators apply for new licenses to operate with LEU fuels, Russian regulatory entities would take a fresh look at the facilities' standards, practices, and objectives. Accordingly, it is plausible that the conversion process could ultimately lead to pressures to shut down rather than convert some Russian facilities.[52]

Indeed, Russia has already made decisions to shut down and/or decommission some facilities on economic grounds. At the 2010 RERTR meeting, Arkhangelskiy tried to argue that several recent shutdowns of critical assemblies demonstrated Russia's commitment to HEU minimization. But in an interview, Arkhangelskiy confirmed that economic considerations weigh heavily in Rosatom's decision-making regarding shutdowns/decommissions:

> When a decision is made about shutting down a reactor … the main issue is the economic issue … The issue of safety and security of HEU use has not been discussed. A great number of these HEU reactors were shut down and many more will be shut down in the future.[53]

Only one economic factor could favor Russia's conversion of its research reactors to LEU fuel. It is possible that Rosatom envisions potential profits from supplying similar LEU-Mo fuel to European markets.

### *Security and Nonproliferation*

Neither Rosatom nor Russian nuclear scientists expressed concern about the security of their facilities. Despite the fact that terrorists attacked Moscow's Domodedovo Airport while the interviews for this chapter were being conducted in the Russian capital, no interviewee even made mention of terrorism. Only the Obninsk official cited physical security and nonproliferation as important goals, but he gave equal priority to "economic efficiency." This largely accords with existing literature's characterization of Russian attitudes toward security. In the absence of specific, recent, known attempts by terrorists or thieves to steal Russian nuclear material, individual institutes and government entities tend to view their own facilities as secure. Ironically, two decades of U.S. assistance to improve physical security at Russian nuclear facilities has contributed to the Russian government's lack of concern with nuclear theft and/or terrorism.

### Decision-Making Authority

Research conducted to determine the inter-organizational dynamics of potential conversion yielded several insights regarding the relationship between Rosatom and individual institutions. When asked to discuss Rosatom's level of influence over conversion decisions at these facilities, Arkhangelskiy prefaced his response by noting that it was "a difficult question."[54] But he added that because "Rosatom enterprises" are responsible for producing fuel for all research reactors, this gives Rosatom a measure of influence over conversion decisions. "From this standpoint the influence of Rosatom in making decisions is very high."

The Machine-Building Plant (MSZ) in Electrostal (Elemash) supplies uranium dioxide-based fuels to several research reactors: the SM-3, BOR-60, and PIK.[55] The Novosibirsk Chemical Concentrates Plant "produces a range of cermet fuels used in MR, VVR, IRT, and IVV type research reactors."[56] The Novosibirsk plant is working on the development of high uranium-density U-Mo alloy fuels that may make it possible to convert high-performance reactors.[57] Both the Elemash and Novosibirsk plants are owned by the TVEL Corporation.

Complicating conversion decision-making authority, however, is the fact that "the operator is under the control of those parties who are responsible for the management of nuclear power, and operator organizations belong to different state organizations."[58] Moreover, "a large number of research reactors" are not under Rosatom's control, leaving conversion decisions "up to these management bodies" of the facilities with research reactors.[59] According to Arkhangelskiy, even where facilities are not subject to Rosatom's authority, Rosatom's influence "is very high" and, furthermore, "if Rosatom makes a decision or recommendation to convert a reactor it will be taken under consideration by the state bodies."[60]

At the same time, Arkhangelskiy acknowledged that "in some cases there can be serious opposition" from operators and/or management bodies, making it difficult to persuade certain reactor facilities to convert.[61] The St. Petersburg University of Nuclear Physics was described as enjoying substantial autonomy.[62] This facility has two reactors using HEU with an enrichment of 90 percent. One, the WWR-M, has been operational since 1959. The other, PIK, is not projected to start operating at power until 2014, but it reportedly achieved criticality in 2011 after many years of delay. This makes it the world's first new HEU-fueled research reactor to go critical in seven years, and only the second in more than two decades.[63] Accordingly, the PIK may represent an especially difficult challenge for conversion, in light of its recent startup and independent management.

Arkhangelskiy's characterization of Rosatom's relationship with St. Petersburg University merits examination:

> Judging from my long discussions with personnel at the University, I think they will oppose very much the conversion of these reactors. However, the situation is evolving and maybe they will start the dialogue and maybe in the future one day they will agree to convert their reactors. But, I repeat, it's up to them to make the decision. And from the formal point-of-view they can disagree with the decision.[64]

Arkhangelskiy described how St. Petersburg is technically under control of Kurchatov Institute, a national research center that includes, along with St. Petersburg, two other institutes. It has been speculated that the PIK reactor's delay in starting operations has resulted from a lack of funds,[65] but Arkhangelskiy hinted that other forces may also be at play. "From the technical point of view there are no obstacles to the start-up of the PIK reactor, and they have the necessary authorization from the regulatory body. Only the decision of the director [will matter] … and it should be politically correct."[66]

The Obninsk official largely confirmed Arkhangelskiy's characterization of Rosatom's influence, describing how "the facilities work usually in frames of the [Rosatom-sponsored research] programs or on [Rosatom's] request."[67] Certain research institutes, however, also have contracts with "facilities inside of Rosenergoatom (NPP or TVEL) or maybe outside Rosatom (commercial activities)." While Rosatom research and development programs provide some funding, commercial contracts for "isotope production, heat and electricity production like RIAR in Dimitrovgrad," also sustain reactor operations. Such funding from alternative sources to these institutes may reduce Rosatom's ability to affect conversion decisions.

The Obninsk official also suggests that the role of operators in conversion decisions depends on the type of research being conducted. If reactors serve a commercial purpose such as producing medical isotopes, or a "facility work[s]" in furtherance of "some special program," institute officials will resist Rosatom's authority and side with operators opposed to conversion. Although the Obninsk official stated that an individual operator "is not [the] main person to make [the] final decision," operators likely play a role in their respective institutes' amenability to conversion.[68]

## Summary of Findings

Rosatom's recent declaration of intent to convert some domestic research reactors to LEU fuel reflects an authentic attitudinal change. There are several contributing factors to Rosatom's new embrace of domestic conversion. Most important was the U.S. government's efforts to convert all of its domestic reactors, and its demonstrated ability to do so in most cases without substantial loss of neutron flux. While other, normative trends may have influenced Russian decision-making, there is little evidence at this time that they were decisive. Such trends could, however, play a role in influencing future Russian policy.

Development of high-density LEU fuel remains a primary technical challenge, particularly with respect to high-performance research reactors, which require the establishment of new manufacturing capabilities and the licensing of new fuel. Moreover, problems have arisen in the qualification of recently developed LEU-Mo fuels, delaying further tests that are a necessary prerequisite to conversion. On the other hand, feasibility studies conducted under the U.S.–Russia Implementing Agreement have concluded that four reactors can be converted, and are expected to find by the end of 2012 that two more can be converted. As a result, Rosatom has indicated its plan to convert one or two of the reactors to LEU fuel by 2014.

While technical challenges remain, particularly with regard to Russia's high-performance research reactors, the economic concerns of Rosatom and individual institutes are an even greater obstacle to conversion at certain facilities. Specifically, the high cost of LEU-Mo is reported as a deterrent to conversion among reactor operators and the managers of their facilities.

## Recommendations

For this universe of reactors, the top U.S. government priority should continue to be providing assistance to the Russian nuclear facilities that are conducting conversion feasibility studies. Washington should also encourage Moscow to conduct joint feasibility studies for additional Russian facilities. Future conversion decisions will depend on the results of these studies.

Considering Arkhangelskiy's repeated and credible claims that Rosatom's interest in conversion was spurred by U.S. success at converting research reactors, Washington should continue pursuing conversion of its remaining domestic HPRRs. The successful conversion of these U.S. reactors could provide the best hope of persuading Russia to convert reactors such as the PIK, which because it is a new HPRR with an autonomous management body, presents a significant political hurdle to conversion, in addition to potential technical challenges.

More generally, Rosatom's domestic conversion policy has opened the door for more direct negotiations between U.S. GTRI/RERTR officials and Russian nuclear research facilities. To the extent possible, U.S. officials should pursue direct negotiations with reactor operators and their appropriate management bodies. In particular, U.S. officials might prioritize RIAR in Dmitrovgrad, which operates five civilian, HEU-fueled reactors, including four that present substantial technical challenges to conversion: the SM-3; RBT-10/2; RBT-6; and the BOR-60.[69]

## Notes

1 Most research reactors operate at relatively low power. Their fuel requirements therefore are not as large as those of reactors that produce electricity or

nuclear-weapons materials. Overall, however, Russian and Soviet research reactors have consumed considerable amounts of enriched uranium. Data on the operating histories of reactors suggest that, by the end of 2009, the research reactors that were designed and built by the Soviet Union had used about 11.3 tons of 90 percent HEU, 1.2 tons of 80%-enriched uranium, and 6 tons of HEU with enrichment of 36 percent. See Pavel Podvig, "Consolidating Fissile Materials in Russia's Nuclear Complex," in *Fourth Annual Report of the International Panel on Fissile Materials*, May 2009: 3; "Global Fissile Material Report: Balancing the Books in 2010," *Fifth Annual Report of the International Panel on Fissile Material*, May 2010: 63; J.E. Matos, "Technical Challenges for Conversion of Civilian Research Reactors in Russia," presentation at the National Academy of Sciences, Washington, D.C., 29 November 2010, http://dels.nas.edu/resources/static-assets/nrsb/miscellaneous/MatosPresentation.pdf (accessed on 27 September 2012).

2 Elena Sokova, "Phasing Out Civilian HEU in Russia," *Nonproliferation Review* 15,2 (2008): 224.

3 Cristina Hansell, "Practical Steps Toward a World Without Civilian HEU," *Nonproliferation Review* 15,2 (2008): 292.

4 Matos, "Technical Challenges," Slide 2.

5 Matos, "Technical Challenges," Slide 4. In 1993, the two states also signed the HEU Purchase Agreement, providing a basis for the United States to purchase HEU from Russia's nuclear weapons program to be converted into LEU and used in U.S. commercial nuclear power reactors. Richard Falkenrath, "The U.S.–Russian HEU Purchase Agreement: Achievements, Problems, Prospects," CSIA Discussion Paper 95–07, Kennedy School of Government, Harvard University, July 1995.

6 Matos, "Technical Challenges," Slide 4. The Bochvar Institute took the lead on Russian fuel development.

7 Material Consolidation and Conversion Program.

8 The Global Initiative to Combat Nuclear Terrorism.

9 Parrish Staples, U.S. Department of Energy, "GTRI and HEU Minimization Efforts," presentation for Global Phase-out of Civilian HEU PRP, LBJ School of Public Affairs, Austin, TX, 26 October 2010.

10 Matos, "Technical Challenges," Slide 4.

11 "U.S., Russian Federation Sign Joint Statement on Reactor Conversion," U.S. Department of Energy, 26 June 2012, http://nnsa.energy.gov/mediaroom/pressreleases/jointstatement062612_(accessed on 27 September 2012), reported that "In the U.S., 20 out of 27 reactors have either been converted to use LEU fuel or shut down."

12 Sokova, "Phasing Out Civilian HEU in Russia," 215.

13 "Russia," Nuclear Threat Initiative, last modified November 2009, http://www.nti.org/db/heu/russia.html (accessed on 1 November 2012).

14 Thomas Young, Cole Harvey, and Ferenc Dalnoki-Veress, "It's Not Just New Start," James Martin Center for Nonproliferation Studies, 21 December 2010, http://cns.miis.edu/stories/101221_nuclear_agreements.htm (accessed on 27 September 2012), citing Ivan Kamenskikh, Deputy Director General of Rosatom.

15 Young et al., "It's Not Just New Start."

16 Ivan Kamenskikh, Deputy Director General, Rosatom, RERTR 2010 Meeting, Lisbon, Portugal, October 2010.

17 "Project #245: Radleg," http://www.kiae.ru/radleg/ch6e.htm (accessed on 10 April 2011). See also Armando Travelli, "RERTR: Origins, Challenges, Progress and Lessons," Powerpoint presentation, LBJ School of Public Affairs, University of Texas at Austin, 5 October 2010, which surveys various discrete Russian reactor designs.

18 Sokova, "Phasing Out Civilian HEU in Russia," 214.

19 Kenley Butler, "Russia: Nuclear Research Facilities," Nuclear Threat Initiative, last modified July 2004, http://www.nti.org/db/nisprofs/russia/reactor/research/with/overview.htm (accessed on 10 April 2011).
20 "U.S., Russian Federation Sign Joint Statement on Reactor Conversion."
21 Matos, "Technical Challenges," Slide 8.
22 Matos, "Technical Challenges," Slide 9.
23 Matos, "Technical Challenges," Slide 14, indicates the following: SM-3, RBT-10/2, RBT-6, PIK.
24 Matos, "Technical Challenges," Slide 18.
25 Staples, "GTRI and HEU Minimization Efforts."
26 The Russian Government mandated the State Atomic Energy Corporation ROSATOM negotiate with the DOE on the signing of an implementing agreement "concerning cooperation in research into feasibility of conversion of the Russian research reactors to use low enriched uranium fuel." Related directive No 1919-r of 30 October 2010 was signed by President Vladimir Putin. "ROSATOM and USA will discuss conversion of Russian research reactors to LEU fuel," *Nuclear.Ru*, 10 November 2010, http://www.rosatom.ru/wps/wcm/connect/rosatom/rosatomsite.eng/presscentre/nuclear_industry/34d1fa0044a2a90e8201def06803e96b (accessed on 27 September 2012).
27 Staples, "GTRI and HEU Minimization Efforts."
28 The Strela at IPPE was previously known to be closed. Arkhangelskiy also indicated the closure of the FS-4, FS-5 at NIKIET/RDIPE.
29 "U.S., Russian Federation Sign Joint Statement on Reactor Conversion." According to a Harvard report, studies on two of the six Russian reactors were completed by the original deadline of December 2011, finding conversion to LEU fuel to be feasible. Matthew Bunn and Eben Harrell, "Consolidation: Thwarting Nuclear Theft," Managing the Atom Program, Harvard University, March 2012, http://belfercenter.hks.harvard.edu/files/Consolidation_Thwarting_Nuclear_Theft_corrected.pdf (accessed 13 November 2012).
30 "Russia to Convert Two Reactors to LEU in 2014," *RIA Novosti*, 26 June 2012, http://en.rian.ru/russia/20120626/174248257.html (accessed on 1 July 2012).
31 The Argus reactor is also being explored as a potential production source for the medical isotope Mo-99, after conversion.
32 "С. Кириенко: Стоимость перевода шести реакторов на НОУ может достичь US$12,5 млн," http://www.nuclear.ru/rus/press/other_news/2126825/, *Nuclear.ru*, 26 June 2012 (accessed on 3 July 2012).
33 Christina Hansell and Anna Loukianova, "Leveraging U.S. Policy for a Global Commitment to HEU Elimination," in *The Global Politics of Combating Nuclear Terrorism: A Supply Side Approach*, eds. William Potter and Christina Hansell (New York: Routledge, 2010), p. 41.
34 Matos, "Technical Challenges," Slide 6, stating, "There are no commitments to conversion and no schedules because feasibility studies have not been completed and no agreements to convert reactors have been made by Rosatom."
35 Pavel Podvig, "Russia to begin conversion of its research reactors to LEU in 2013," http://fissilematerials.org/blog/2012/10/russia_to_begin_conversio.html (accessed on 13 November 2012).
36 Inter alia, the resolution "Calls upon all States to manage responsibly and minimize to the greatest extent that is technically and economically feasible the use of highly enriched uranium for civilian purposes, including by working to convert research reactors and radioisotope production processes to the use of low enriched uranium fuels and targets."
37 U.S. government official's correspondence with the author.
38 Nikolai Arkhangelskiy, interview with author (interpreter's translation from Russian to English), Moscow, Russian Federation, 26 January 2011.

39 Arkhangelskiy noted that "15 years ago it was practically impossible to convert high-powered reactors; the developments of U-Mo fuel were launched and they were initiated by the Americans."
40 Dr. Dale Klein, interview with author, Austin, Texas, 4 March 2011.
41 Kiriyenko said that, following detailed studies, the conversion of Russian research reactors from HEU to LEU fuel will be made where it is economically and technically justified. He said that if the results of the studies show that the conversion was feasible, the actual conversion would be carried out under a separate agreement. "Russian research reactors may convert to LEU," *World Nuclear News*, 10 December 2010, http://www.rosatom.ru/wps/wcm/connect/rosatom/rosatomsite.eng/presscentre/nuclear_industry/3796658044fe1863abdeabe0d43de87e (accessed on 2 July 2012).
42 Arkhangelskiy specifically said LEU-Mo fuel had not been "qualified in a proper manner."
43 Matos, "Technical Challenges."
44 A. Izhutov, V. Alexandrov, A. Novosyolov, "The Main Results of Investigation of Modified Dispersion LEU-Mo Fuel Tested in the Mir Reactor," RERTR 2010: 32nd International Meeting on Reduced Enrichment for Research and Test Reactors, Lisbon, Portugal, 10 October 2010.
45 O.A. Golosov, S.A. Averin, V. L. Panchenko, M.S. Lyutikova, "An Influence of Porosity Development in U-Mo Fuel Particles on the Performance of U-Mo/Al Dispersion Fuel," RERTR 2010: 32nd International Meeting on Reduced Enrichment for Research and Test Reactors, Lisbon, Portugal, 10 October 2010.
46 Arkhangelskiy, interview with author.
47 Arkhangelskiy, interview with author.
48 Yavgeni Suderov, discussion with the author, Moscow, Russian Federation, 25 January 2011.
49 Gennadi Pshakin, email message to Paul Osborne, 20 March 2011. Pshakin also said, "Or conversion process too expensive and facility has no funds for the conversion and Rosatom also has not enough funds allocated for the conversion."
50 Gennadi Pshakin, email message to Paul Osborne, 20 March 2011.
51 U.S. government official's correspondence with the author.
52 U.S. government official's correspondence with the author.
53 Arkhangelskiy, interview with author.
54 Arkhangelskiy, interview with author.
55 Podvig, "Consolidating Fissile Material."
56 Podvig, "Consolidating Fissile Material."
57 Podvig, "Consolidating Fissile Material."
58 Arkhangelskiy, interview with author.
59 Arkhangelskiy, interview with author.
60 Arkhangelskiy, interview with author.
61 Arkhangelskiy, interview with author.
62 Arkhangelskiy, interview with author.
63 "Shtrombakh: First power of PIK reactor expected in 2014," *Nuclear.Ru*, 22 June 2012, http://www.rosatom.ru/wps/wcm/connect/rosatom/rosatomsite.eng/presscentre/announcements/bd93f2004bb52eeab06aba0fd1e28404 (accessed on 27 September 2012); According to Pavel Podvig, "More delays for PIK reactor," *FissileMaterials.org*, 9 August 2011, the "physical start-up" was reported to take place in February 2011, http://www.fissilematerials.org/blog/2011/08/more_delays_for_pik_react.html (accessed on 27 September 2012). The last such HEU-fueled, research reactor to go critical was Germany's FRM-II in 2004 (see chapter by Kenneth Dayman in this book). Preceding that was China's MJTR in 1991, http://www-naweb.iaea.org/napc/physics/research_reactors/database/rr%20data%20base/datasets/report/China,%20People's%20Republic%20of%20%20Research%20Reactor%20Details%20-%20MJTR.htm (accessed on 27 September 2012).

64 Arkhangelskiy, interview with author.

65 Frank von Hippel. "Global Cleanout of Highly-enriched Uranium," (presentation at the University of Texas, Austin, Texas, 19 October 2010).

66 Arkhangelskiy, interview with author.

67 Gennadi Pshakin, email message to Paul Osborne, 20 March 2011.

68 Gennadi Pshakin, email message to Paul Osborne, 20 March 2011.

69 Matos, "Technical Challenges," Slides 18-22. Although the BOR-60 fast reactor is scheduled to be shut down in 2015, it is "to be replaced by 2019 by MBIR, a new 100 MW sodium cooled fast reactor...[with] fuel [which] is likely to be MOX (UO2 – PuO2) or nitride (UN – PuN), probably with HEU."

# 11 Russia

## Critical Assemblies and Pulsed Reactors

*Paul Osborne*

Critical assemblies and pulsed reactors are a relatively new focus among those seeking to minimize the worldwide non-weapon use of highly enriched uranium (HEU), in order to reduce risks of nuclear proliferation and nuclear terrorism. These facilities have lifetime cores, which makes them harder economically and politically to convert to low enriched uranium (LEU) fuel that is unsuitable for nuclear weapons. The original strategy of the Reduced Enrichment for Research and Test Reactors (RERTR) program – converting to LEU during refueling – is not applicable because these facilities have no need, budget, or plan to refuel. Yet, these facilities remain highly attractive targets for those seeking to make nuclear weapons because they contain large amounts of very lightly irradiated HEU. If a state or terrorist obtained a relatively small amount of this material, they could likely build a crude nuclear weapon using information available through open sources, including the internet.[1]

In light of the dangers of nuclear terrorism and proliferation, policymakers should focus attention and resources on the possibilities for conversion, decommissioning, or consolidation of critical assemblies and pulsed reactors. Russia alone is home to 16 operating, HEU-fueled pulsed reactors and more than half of the world's HEU-fueled critical assemblies (see Table 11.2).[2] Accordingly this chapter focuses on the prospects of converting to LEU, consolidating, or closing Russia's pulsed reactors and critical assemblies. As will be shown, the challenges to phasing out such use of HEU are complex, entailing technical and economic obstacles, and domestic and international politics.

Economic concerns are the most common obstacle expressed by Russian officials to converting or decommissioning these facilities. The costs of manufacturing exotic, high-density LEU fuels, as well as the costs associated with converting or decommissioning, play a large role in their calculations. But they also have technical concerns, particularly related to the continuing development of suitable high-density LEU fuels that could be used in the conversion of facilities that already have compact cores, such as pulsed reactors.[3]

Human nature and the narrow interests of workers also can be obstacles. Concerns about salary decreases, job cuts, and the loss of status as an HEU-fueled facility are all very real among workers at these facilities. Such resistance to change may also be partly an historical legacy, as many of the facilities in question were built – and many of the current operators worked at them – during the Soviet era. The Soviet Union was notoriously resistant to change, and this mentality may partly explain the reluctance of managers and operators to convert or shut facilities that they deem to be functioning as well as ever.[4]

But there are also some hopeful signs in U.S.–Russian cooperation. The U.S. decision to convert all of its research reactors to LEU was not ignored in Russia, and has influenced decisions by Russian officials to convert some of their own reactors to LEU.[5] The December 2010 U.S–Russia agreement, to conduct feasibility studies on converting six Russian research reactors, suggests that a new level of bilateral cooperation to minimize non-weapons uses of HEU may be possible.[6]

This chapter analyzes what is known about the prospect of converting Russian critical assemblies and pulsed reactors to LEU or decommissioning them. It starts by summarizing the findings of previous literature, next explains the methodology used for this chapter's new research, and then presents the findings from interviews conducted with Russian officials. The chapter concludes with policy recommendations and suggestions for future research.

## Literature Review

Critical assemblies and pulsed reactors are a special challenge for global efforts to phase out non-weapons uses of HEU. Their lifetime cores do not allow for conversion during refueling because they do not require fresh fuel. Although conversion is still an option, operators have expressed concern about the downtime associated with converting their functioning facilities. Operators typically feel that they have no incentive to convert.[7]

These facilities have lifetime cores because they are "low power" or "zero power" reactors. In contrast to "steady-state" research reactors, they operate only intermittently. Accordingly, they have low fuel burn-up rates, leading to only low levels of fission-product buildup in the fuel, and thus low levels of radioactivity. The resulting low risk of radiation poisoning makes theft from their HEU cores attractive for those who aspire to construct a nuclear weapon.[8]

Pulsed reactors were designed to mimic the neutron bursts from a nuclear -weapon explosion in order to determine the effects on nuclear warheads and other objects.[9] The reactors produce a pulse that only lasts for a short time, which allows for a higher density neutron flux than could be achieved from the same reactor operating at a continuous state. In order to produce a flux with the desired level of neutron density, large quantities of HEU are used.[10] Most notably, the BIGR pulsed reactor, at the All-Russian Scientific Research Institute of Experimental Physics (VNIIEF) in Sarov, contains 833 kg of 90%-enriched uranium in its core. Although the BIGR is the largest, Russia has a total of 19 pulsed reactors that collectively contain two tons of HEU in their cores.[11]

While it has been noted that the compact cores and high levels of enrichment make conversion of pulsed reactors difficult, a relatively optimistic proposal was presented by VNIIEF at the 2005 RERTR conference, exploring the possibility of converting their six HEU-powered, pulsed reactors – including the BIGR – to LEU fuel.[12] The VNIIEF authors suggest that two of these reactors could be converted to LEU with existing fuels – uranyl sulphate and uranium-graphite – because their cores are equipped with neutron moderators. By contrast, in the case of the four un-moderated reactors, they find that the increase in weight from LEU fuel might require designing new, replacement reactors. In all cases, however, the VNIIEF paper suggests that LEU-fueled reactors could provide similar or superior levels of performance compared to the current HEU versions.[13]

Russia's need for so many pulsed reactors has also been questioned. Today, the effects of neutron bursts can be modeled by computer simulations. In the United States, this rationale was used to justify the announced shutdown of one of the two HEU-fueled pulsed reactors at Sandia National Laboratory.[14] The U.S. laboratory did not, however, explain the justification for continuing to operate the second.

Critical assemblies are mockups of reactors. They are "zero-powered" and thus do not require the same kind of cooling systems as typical test and research reactors. The assemblies can be quickly reconfigured and are used primarily for physics experiments as well as modeling properties of proposed reactor cores.[15] Many critical assembly cores contain HEU. Russia is home to more than half of the world's approximately 50 HEU-fueled critical assemblies. The amount of HEU in critical assemblies varies depending on their function, as illustrated in Table 11.1.[16]

Table 11.2 is a chart listing Russia's operating HEU-fueled critical assemblies and pulsed reactors, as of 2008. This table includes the specific locations of reactors, where available, and other updates to a previously published list by Reistad and Hustveit.[17]

Converting critical assemblies to LEU is feasible, but only after the actual reactor converts. Until that point, mockups of HEU-fueled reactors require HEU. The unnecessary risk in Russia, however, is the use of 90%-enriched uranium in the mockups of reactors that do not use fuel with that high of an enrichment level. For example, Russian naval reactors use uranium enriched to approximately 40 percent, on average, and fast-neutron reactors use uranium that is enriched to less than 30 percent. Still, the critical assemblies associated with these reactors use 90%-enriched HEU, in combination with "medium-enriched" HEU and depleted uranium (having less than the 0.7% enrichment of natural uranium). Russian officials have argued that using this combination, including 90%-enriched HEU, allows for fine-tuning of the fuel in the assembly. But such fine-tuning alternatively

*Table 11.1* Critical Assembly Types

| *Mockup Type* | *Quantity of HEU in Core* |
|---|---|
| Research Reactor | A few kg |
| Space Reactor | From tens to over 100 kg |
| Naval Reactor | More than 100kg |
| Fast-Neutron Power Reactor | Ton-quantities |

*Table 11.2* Currently Active HEU-Fueled Critical Assemblies and Pulsed Reactors in Russia

| *Reactor* | *Type* | *Location* | *Status* |
|---|---|---|---|
| 659 | Critical Assembly | | Active |
| 1125 | Critical Assembly | | Active |
| AKSAMIT | Critical Assembly | | Active |
| Astra | Critical Assembly | | Active |
| BARS-1 | Pulsed | Sergiyev Posad | Active |
| BARS-4 | Pulsed | Lytkarino - NIIP | Active |
| BARS-5 | Pulsed | Snezhinsk - VNIITF | Active |
| BARS-6 | Pulsed | Lytkarino - NIIP | Active |
| BFS-1 | Critical Assembly | Obninsk - IPPE | Active |
| BFS-2 | Critical Assembly | Obninsk - IPPE | Active |
| BIGR | Pulsed | Sarov - VNIIEF | Active |
| BIR-2M | Pulsed | Sarov – VNIIEF | Active |
| BR-1 | Critical Assembly | Sarov - VNIIEF | Active |
| BR-K1 | Pulsed | | Active |
| CA MIR.M1 | Critical Assembly | | Active |
| CA-SM | Critical Assembly | | Active |
| Delta | Critical Assembly | | Active |
| Emphir-2M / EFIR-2M | Critical Assembly | | Active |
| EFIR-2M | Critical Assembly | Moscow - Kurchatov | Active |
| FBR-L | Pulsed | Snezhinsk - VNIITF | Active |
| FKBN-2M | Critical Assembly | | Active |
| FM MR | Critical Assembly | | Active |
| FS-1M | Critical Assembly | | Active |
| GIR 2 | Pulsed | Sarov - VNIIEF | Active |
| HYDRA (IIN-3M Gidra) | Pulsed | Moscow - Kurchatov | Active |
| Igrik | Pulsed | Snezhinsk - VNIITF | Active |
| Impuls 1 Mil. | Pulsed | | Active |
| Impuls 2 Mil. | Pulsed | | Active |
| Impuls 6 Mil. | Pulsed | Sergiyev Posad | Active |
| ISKRA | Critical Assembly | | Active |
| KV-1 | Critical Assembly | | Active |
| Kvant | Critical Assembly | | Active |
| MAKET | Critical Assembly | | Active |
| Nartsiss-M2 | Critical Assembly | | Active |
| PhS-2 (FS-2) | Critical Assembly | Moscow - NIKIET | Active |
| PhS-4 (FS-4) | Critical Assembly | Moscow - NIKIET | Shut down announced |
| PhS-5 (FS-5) | Critical Assembly | Moscow - NIKIET | Shut down announced |
| PIK FM | Critical Assembly | Gatchina - PNPI | Active |
| RF-GS | Critical Assembly | | Active |
| SF-1 | Critical Assembly | | Active |
| SF-7 | Critical Assembly | | Active |
| SGO | Critical Assembly | | Active |
| SO-2M SubCA | Critical Assembly | Moscow - NIKIET | Active |
| STEND-2 | Critical Assembly | | Active |
| STEND-3 | Critical Assembly | | Active |

*Table 11.2* continued

| *Reactor* | *Type* | *Location* | *Status* |
|---|---|---|---|
| STEND-4 | Critical Assembly | | Active |
| STEND-5 | Critical Assembly | | Active |
| Strela | Critical Assembly | Obninsk - IPPE | Shut down announced |
| T2 | Critical Assembly | | Active |
| VIR-2M | Pulsed | Sarov - VNIIEF | Active |
| Yaguar (NHUAR) | Pulsed | Snezhinsk - VNIITF | Active |

Key:
IPPE = Institute of Physics and Power Engineering
JINR = Joint Institute of Nuclear Research
NIIP = Scientific Research Institute for Instruments
NIKIET = Research and Development Institute of Power Engineering
PNPI = Petersburg Nuclear Physics Institute
VNIITF = All-Russian Scientific Research Institute of Technical Physics

Notes:
IBR-30, pulsed reactor, at Dubna – JINR, which is on some lists, was reportedly shut down in 2001.
BR-1, at Obninsk-IPPE, was reportedly shut down in 2010. (According to Pavel Podvig, email to Alan J. Kuperman, 12 June 2012.)

could be achieved by using only medium-enriched HEU and depleted-uranium, avoiding entirely the use of 90%-enriched HEU that presents the greatest risks of nuclear terrorism and proliferation.[18]

Decommissioning is another viable option for most of Russia's critical assemblies. As with pulsed reactors, most critical assembly functions can be performed with computer simulations. Many of the "benchmark experiments" have already been performed by critical assemblies and their results made available to the scientific community. In fact, 38 research reactors have been converted from HEU to LEU by the RERTR program without the aid of critical assemblies.[19] The International Criticality Safety Benchmark Evaluation Project has dedicated itself to compiling data from around 3,800 configurations and distributing the evaluations to 60 countries.[20]

For the few applications in which HEU-fueled critical assemblies and pulsed reactors may still be necessary, consolidation to fewer facilities may be useful. Previous advocacy of such consolidation has focused on critical assemblies, but the same principle applies to pulsed reactors. In Europe, past consolidation to just a few HEU-fueled critical assemblies has been effective, and some argue that even further consolidation is feasible. For example, Frank von Hippel argues that of the five remaining HEU-fueled critical assemblies in Europe, only the two mockups of fast-neutron reactors need to operate on HEU. One of these two, the Masurca, uses fuel enriched to only 30 percent, leaving only one that requires very high enrichment, according to von Hippel. Assuming this expert is correct that all of Europe requires only one critical assembly fueled with 90%-enriched HEU, then it is highly unlikely that Russia needs anywhere close to 30 of them.[21]

Previous literature also has explored some potential incentives for conversion, decommission, or consolidation of these Russian facilities. For example, the HEU in the cores of Russia's critical assemblies and pulsed reactors could be blended down into LEU for commercial sale or use as fuel in nuclear power plants. (Admittedly, existing programs blend down only fresh HEU, but they could be expanded to incorporate lightly irradiated HEU, as discussed below.) Each metric ton of 90%-enriched uranium could be used to create roughly 25 tons of 4.3%-enriched uranium, valued at about $25 million in 2004.[22] By 2011, the value of such fuel had increased nearly three times.[23]

If some of these facilities were decommissioned, researchers could still perform experiments through cooperation at other facilities. For example, the Generation-IV International Forum established a framework for international cooperation in developing advanced reactor systems, few of which are expected to require HEU. The mockup assemblies for such reactors could be internationally accessible and minimize the use of HEU. Russia and other countries that decommission HEU-fueled critical assemblies and pulsed reactors would also reap financial rewards from no longer having to provide the highest level of security at these sites, which can be extremely expensive.[24]

All of this publicly available information suggests that it would be in Russia's security and financial interest to convert most or all of its pulsed reactors and critical assemblies to LEU or to decommission them in favor of a consolidation arrangement. But Russia historically has been hesitant to make such changes. The literature offers several possible explanations for this reluctance: the short-term financial cost of converting to LEU; persistent underestimating of the threat of nuclear terrorism; overconfidence that recent security upgrades can guarantee protection; and the fear of losing a technological advantage.[25]

But recent years have begun to show progress. In 2010, Russia announced the shutdown of one critical facility, the Strela at the Institute of Physics and Power Engineering (IPPE) in Obninsk, and two subcritical facilities,[26] the FS-4 and FS-5 at the Research and Development Institute of Power Engineering (NIKIET) in Moscow.[27] These announcements were the first indication of a change in Russia's attitude towards phasing out non-weapons HEU use on its own territory. Since then, Russia also has conducted feasibility studies on converting six of its steady-state research reactors from HEU to LEU fuel, and announced plans to convert at least two of them, starting in 2014,[28] which suggests that Moscow finally has embraced a policy of domestic HEU minimization.

Although the existing literature is informative, it fails to resolve certain questions. For example, there is little detail on the specific reasons that the Russian government traditionally has avoided conversion, decommission, or consolidation. It is unclear why Russia values so highly these HEU-fueled, low- or no-power nuclear facilities, when other states and regions have very few of them. There is also sparse reporting about past and prospective cooperative efforts between the United States and Russia to reduce HEU use at such facilities.

## Methodology

The new research conducted for this chapter aimed to address several questions:

- Why has Russia avoided converting, consolidating, or decommissioning critical assemblies and pulsed reactors up to this point? What was the value of such HEU-fueled facilities in the past, and is there any reconsideration?
- In light of documented thefts of HEU in Russia and ongoing terrorist threats, does the Russian government see incentive for HEU minimization?
- Has the United States encouraged Russia to convert such facilities? When and how? Has this affected Russia's thinking on the matter? What kinds of external pressure and support have helped and hurt the process? Are there any additional incentives that could be used to promote the further reduction of the number of such HEU-fueled facilities?
- Why has Russia recently decided to cooperate on HEU minimization within Russia? What is the scope of its plan, and what are the future intentions for conversion or decommission of nuclear facilities?

I investigated these questions through primary sources, discussing critical assemblies and pulsed reactors with Gennadi Pshakin and Yevgeny Sidirov from IPPE in Obninsk, near Moscow, and Nikolay Arkhangelskiy from Rosatom in Moscow. Interviews took place in several formats, including in-person in Moscow, via Skype conversations, or through email correspondence.

Although these interviews were extremely useful, the primary research also confronted obstacles. The conversion of Russian critical assemblies and pulsed reactors is a relatively narrow topic, so there is only a short list of potential interviewees. Simply establishing contact with the officials that I wished to interview was a challenge. Many of my email requests went unanswered, and some officials responded only to inform me that they did not have the time or desire to speak to me. That said, those who did engage with me were very helpful.

## Findings

The main obstacles to conversion or decommission of Russia's critical facilities and pulsed reactors appear to be economic. First, and most obviously, converting or decommissioning a facility costs a substantial amount of money. This economic obstacle appears to be the primary reason that Russia still has so many of these HEU-fueled facilities. Historically, funds were not allocated in advance to decommission a critical assembly after its intended purpose was served, and Russian officials have not been eager to dedicate scarce resources to ending past projects. Often, critical assemblies ready for decommission are instead incorporated into new programs, effectively giving them "new life." Still, Dr. Pshakin insists there has been some progress. For example, since 1994, he claims the number of nuclear facilities at the IPPE in Obninsk has been cut in half, from 34 to 17.[29]

One cost of decommissioning critical assemblies or pulsed reactors arises from the fact that the fuel extracted from the core is slightly irradiated. Storage of irradiated fuel is expensive, and there has been no alternative because the original, U.S.–Russia program to blend Russian HEU into LEU for power reactors accepted only fresh HEU. That bilateral program to blend down of hundreds of tons of fresh Russian HEU into LEU fuel is an important accomplishment, but failed to address the substantial amount of irradiated HEU fuel that has accumulated in Russia during 60 years of nuclear research.[30] Reportedly, Russia is now exploring the blend-down of slightly irradiated HEU into LEU fuel, and the reprocessing of more highly irradiated HEU, and the United States may provide financial assistance.

As for the conversion option, it would require purchase of new fuel by facilities with lifetime cores that had no expectation of, nor budget for, such an expense. Moreover, high-density LEU fuel costs more than HEU fuel, at least at the moment, thereby increasing the expense. Still, Russia is apparently working on a new high-density LEU fuel it expects would allow for conversion of additional facilities,[31] and the economies of scale from increased production could eventually reduce costs.

When assessing the prospect of conversion or decommission, Russian analysts conduct a cost-benefit analysis, according to at least one informed source. They assess the probability of a security incident at that facility and its probable cost, and compare that to the costs of conversion or decommission. This analysis reportedly determines whether a facility is converted, decommissioned, or left alone.[32]

The personal interest of employees also can be an obstacle to conversion or decommission. Managers and operators may have economic and psychological interest in keeping a facility operating and HEU-fueled. These personnel may be worried about losing the status associated with an HEU-fueled facility.[33] Given that all facilities cannot be converted at the same time, the operators of those converted first might feel relegated to second-class status. Operators might also be concerned that conversion to LEU would decrease their salaries. For example, radiation status is directly related to salary levels at critical assembly facilities. In fact, if nuclear materials were simply removed from a facility, as would occur in decommissioning, personnel salaries could be reduced by almost half. As a result, even though several Russian critical assemblies are not particularly useful for research at the moment, their operators resist decommissioning them in hopes of finding new uses in the future.[34]

While such concerns are understandable motivation to resist conversion or decommission, the facility operators and managers do not have full control over those decisions. They have more leverage to block such changes if their facility is being used intensively for research purposes or for some other kind of special program that the government wants to continue uninterrupted. They also have greater ability to oppose conversion if Rosatom lacks the funds to pay for it.[35]

But Dr. Pshakin insists that, in general, Rosatom has substantial control over these nuclear facilities. Critical assemblies appear to operate within a program created by or at the request of Rosatom. They may also have contracts within

Rosenergoatom, which operates Russia's nuclear power plants.[36] But most of Russia's nuclear facilities that do not operate under commercial contracts are funded by Rosatom R&D programs, Pshakin said.[37] Interestingly, Arkhangelskiy disputes this viewpoint, claiming that many such facilities are not directly under Rosatom control.[38]

One technical constraint on conversion or decommission is that some critical assemblies are used to model reactors that still use HEU fuel. To properly simulate such reactors, the assembly must use HEU as well. This is particularly an issue for critical assemblies that model fast reactors, such as the BFS1 and BFS2 reactors at the IPPE in Obninsk.[39] These critical assemblies can only be converted to LEU if and when the fast research reactors they model are converted to LEU, which may not prove feasible. But these reactors typically use a medium-enriched HEU, so in the meantime it is advisable and technically feasible to reduce the enrichment of uranium at the critical assembly to no higher than that in the reactor.

Pulsed reactors pose unique challenges to conversion. Most of Russia's pulsed reactors are in the defense sector, and interviewees declined to discuss those. But some pulsed reactors are in the civilian sector, such as the Hydra reactor at the Kurchatov Institute in Moscow, which uses a uranium-sulfate HEU fuel, enriched to 90 percent. The pulsed reactors typically have compact cores, which makes conversion harder. There is concern that the volume increase associated with conversion to LEU could inhibit the ability of the reactor to generate good neutron pulses. That may be one reason that Russia appears not yet ready to consider conversion of even the civilian pulsed reactors.

Arkhangelskiy said that conversion of non-defense pulsed reactors could possibly be considered in the future, after a suitable high-density LEU fuel were created. Indeed, he suggested that LEU fuel already under development, which he said was two or three years from fruition, might prove feasible for converting such reactors. But, in any case, Arkhangelskiy does not expect these facilities to be converted.[40]

As for the prospect of using computer simulations to replace critical assemblies or pulsed reactors, Russian scientists were skeptical. They say that, in Russia, nuclear technologies are more complex than computer models, so experimental validation with reactors is necessary.[41]

On an encouraging note, it seems the United States has had a positive impact on conversion efforts in Russia. In the past, Russia did not consider converting domestic facilities. But the U.S. decision in the last decade that it would convert all of its civilian research reactors to LEU fuel influenced Russian officials. The announcement helped change minds in Russia and encourage domestic conversion. Russian officials say their government takes very seriously the threat of nuclear terrorism, which is why its official policy is to minimize HEU use, including by converting to LEU. In fact, Nikolay Arkhangelskiy says he believes Russia is heading in a direction in which HEU will no longer be used in the future. He claims that there are no real technical benefits to using HEU over LEU, although he acknowledges the temporary economic benefits cited above.[42]

The 2010 U.S.–Russia agreement, under which feasibility studies have been conducted on converting six Russian steady-state research reactors to LEU fuel, provides another reason for optimism. The agreement has enabled new levels of nuclear cooperation between the United States and Russia, including in the civilian sector. A Russian official said that this agreement lays the groundwork for broader cooperation and more concrete steps towards HEU minimization.[43]

Nevertheless, the previously mentioned obstacles to conversion and decommission persist. Rosatom may not prioritize nuclear security and nonproliferation sufficiently to fund conversion and decommission of HEU-fueled facilities. Until their economic concerns are addressed, Russian facility operators will attempt to find new uses for apparently defunct, HEU-fueled facilities to prevent them from being decommissioned. Accordingly, some Russian facility operators express hope that U.S. funding will be made available to help overcome these obstacles.[44]

## Summary of Findings

Economics was the most common obstacle expressed by Russian officials to potentially converting or decommissioning critical assemblies and pulsed reactors. In addition to the costs associated with conversion and decommission, the cost of high-density LEU fuels was also seen as an obstacle. These LEU fuels are currently considerably more expensive than their HEU counterparts. While the price will likely fall in the future, it currently is viewed as a drain on available resources. There are also cost issues concerning disposal of the lightly irradiated HEU fuel that would come from converting or decommissioning functioning critical assemblies or pulsed reactors. Storage of the lightly irradiated fuel is costly, and the lightly irradiated fuel cannot be blended down into LEU as can fresh HEU, under the original U.S.–Russia blend-down program.

Local-level concerns are also a problem. Managers and operators have incentive to keep their facilities functioning on HEU. The loss of radiation status from converting to LEU could trigger salary decreases. Facility managers also worry that decommissioning could lead to loss of jobs. But in all or most cases, Rosatom appears to have the final say over such decisions.

Critical assemblies that model reactors that still use HEU fuel, such as fast reactors, represent a special challenge, but the enrichment of fuel in the assembly can and should be reduced to that of the fuel in the reactor. Pulsed reactors pose a technical challenge because they have compact cores with dense HEU fuel, so conversion would require development of very high-density LEU fuel; even if such fuel is developed, Russia may be reluctant to convert pulsed reactors, especially those in the defense sector. Russia's scientists are skeptical that computer simulations can replace critical assemblies and pulsed reactors.

The most encouraging news is that Russia was positively influenced by the U.S. decision last decade to convert its research reactors to LEU, and as a result has embraced an analogous policy. The 2010 U.S.–Russia agreement to study the feasibility of converting Russian steady-state reactors thus lays the groundwork for broader bilateral cooperation and HEU minimization efforts.

## Conclusion

This chapter's finding that U.S. domestic actions directly influenced Russia's decision to convert its own domestic reactors is important. It signifies that the United States has led by example in phasing out non-weapons use of HEU, and hopefully can continue to do so. U.S. officials should build on recent bilateral efforts to convert Russia's steady-state reactors by expanding the scope to include critical assemblies and pulsed reactors. It is also worth exploring a mutual phase-out of HEU fuel in naval propulsion reactors, as discussed in this book's chapters by Ward and Primachenko.

Future research should aim to verify the findings and expand the scope of this chapter. For example, it would be helpful to interview officials at other Russian civilian critical assemblies and pulsed reactors to determine if they confirm the concerns expressed by those interviewed for this chapter. Although it is a highly sensitive topic, researchers should also explore defense-related pulsed reactors in Russia, and the government's willingness to consider conversion, consolidation, or closure of such facilities.

### *Policy Recommendations*

- Rosatom should agree to not decrease salaries of managers and operators at facilities that convert or are decommissioned. Knowing that their salaries are secure should alleviate some of the fears that personnel may have about conversion or decommission.
- The U.S. government should work with Russia on the conversion or decommission of HEU-fueled critical assemblies, focusing initially on those that do not currently require such fuel. The United States should explore absorbing some of the costs to facilitate conversion, in the interests of U.S. and international security.
- The United States should continue working with Russia to assist in the development of new high-density LEU fuels. The assistance could take the form of sharing knowledge and technical expertise rather than providing funding. Successful development of high-density fuel would allow the United States to encourage Russia to convert its pulsed reactors to LEU.
- The United States and Russia should cooperate in improving computer codes to better simulate the operations of Russian HEU-fueled critical assemblies and pulsed reactors, which could enable closure or consolidation of many such facilities and thereby reduce risks of nuclear terrorism and proliferation.

## Notes

1 Frank von Hippel, "The Need to Address the Larger Universe of HEU-Fueled Reactors, Including: Critical Assemblies, Pulsed Reactors and Propulsion Reactors," paper presented at the 2004 International RERTR Meeting, Vienna, Austria, 7–12 November 2004.
2 Although its quantification of facilities is slightly outdated, an excellent source is Frank von Hippel, "HEU in Critical Assemblies, Pulsed Reactors and Propulsion Systems," paper presented at the Technical Workshop On HEU Elimination, Oslo, Norway, 17–18 June 2006.
3 Nikolay Arkhangelskiy, interview by author. Moscow, Russia, 26 January 2011.
4 Gennadi Pshakin, interviews via email correspondence, 2 January – 28 March 2011.
5 Arkhangelskiy, interview.
6 Pshakin, interviews.
7 Frank von Hippel, "Future Needs For HEU-Fueled Critical Assemblies," paper presented at the Future use of Critical and Subcritical Assemblies, Vienna, Austria, 7–10 February 2005.
8 von Hippel, "HEU in Critical Assemblies."
9 Alexander Glaser and Frank von Hippel, "Global Cleanout: Reducing the Threat of HEU-Fueled Nuclear Terrorism," *Arms Control Today* 22 (2006): 18–23.
10 NTI, "Civilian Uses of HEU." http://www.nti.org/db/heu/civilian.html (accessed on 27 September 2012).
11 von Hippel, "HEU in Critical Assemblies."
12 Vitali Dubinin, "Goals and Objectives of ISTC Project #3128: 'Using Low-Enrichment Uranium in VNIIEF Pulsed Nuclear Reactors,'" paper presented at the 2005 International RERTR Meeting, Boston, Massachusetts, 6–10 November 2005.
13 Dubinin,"Goals and Objectives."
14 Glaser and von Hippel, "Global Cleanout,"18–23.
15 NTI, "Civilian Uses of HEU."
16 von Hippel, "HEU in Critical Assemblies."
17 Ole Reistad and Styrkaar Hustveit, "HEU Fuel Cycle Inventories and Progress on Global Minimization," *The Nonproliferation Review* 15 (July 2008): 266–287.
18 von Hippel, "Future Needs For HEU-Fueled."
19 von Hippel, "Future Needs For HEU-Fueled."
20 Harold F. McFarlane, "Is It Time to Consider Global Sharing of Integral Physics Data?" *Journal of Nuclear Science and Technology* 44 (2007): 1–4.
21 von Hippel, "HEU in Critical Assemblies."
22 von Hippel, "The Need to Address."
23 "The Economics of Nuclear Power," World Nuclear Association, July 2012, http://www.world-nuclear.org/info/inf02.html (accessed on 27 September 2012). It cites a cost estimate from March 2011 of $2,770/kg LEU. This is equivalent to about $2.8 million/ton, or $70 million for 25 tons of LEU.
24 McFarlane, "Is it Time to Consider."
25 Elena K. Sokova, "Phasing out Civilian HEU in Russia," *The Nonproliferation Review* 15 (July 2008): 209–236.
26 "Subcritical" means the fuel is arranged so that the chain reaction is not entirely self-sustaining.
27 Parrish Staples, DOE, presentation at University of Texas at Austin, 26 October 2010.
28 "Russia to Convert Two Reactors to LEU in 2014," *RIA Novosti*, 26 June 2012, http://en.rian.ru/russia/20120626/174248257.html (accessed on 1 July 2012).
29 Pshakin, interviews.
30 Pshakin, interviews.
31 Arkhangelskiy, interview.
32 Yevgeny Sidirov, interview via Skype, Moscow, Russia, 25 January 2011.

33 Pshakin, interviews.
34 Arkhangelskiy, interview.
35 Pshakin, interviews.
36 "Russia: Rosenergoatom," http://www.nti.org/db/nisprofs/russia/govt/rosenerg.htm (accessed on 27 September 2012).
37 Pshakin, interviews.
38 Arkhangelskiy, interview.
39 Arkhangelskiy, interview.
40 Arkhangelskiy, interview. Russia also has one reactor that generates pulses with plutonium fuel, the IBR-2. That reactor recently completed a modernization program, and there are no plans to phase out plutonium fuel in the future.
41 Sidirov, interview.
42 Arkhangelskiy, interview.
43 Pshakin, interviews.
44 Pshakin, interviews.

# Part V

# Military Challenges

# 12 USA and France

## Naval Propulsion

*Rebecca Ward*

The world's first nuclear-powered submarine, the U.S.S. Nautilus, entered service in 1955, ushering in an era of fast-moving submarines capable of remaining at sea for extended periods.[1] Today the United States has the world's largest nuclear-powered fleet with 84 submarines and aircraft carriers, all of which use highly enriched uranium (HEU) fuel, which Frank von Hippel estimates requires the use of approximately 2 tons of HEU per year.[2] Globally, Reistad and Hustveit estimate that naval propulsion requires 3,100 kg of HEU annually, constituting the largest non-weapons use of HEU.[3] HEU was chosen for this application because of its ability to produce a large amount of power in a small volume, making it ideal for space-constrained naval reactor applications. However, its properties also make it an attractive target for both state and non-state actors wishing to develop a crude nuclear weapon. This chapter explores the technical and political prospects of converting the United States' naval reactor fleet from HEU to low-enriched uranium (LEU) in light of the French Navy's progress on conversion. The French naval fleet historically used HEU-fueled reactors, but currently operates all LEU-fueled reactors.

The topic of naval reactor conversion is of particular interest because it represents a military instance of an otherwise civilian problem. As such, the issue must be examined with attention to different priorities and faces greater obstacles than the largely successful conversion of civilian research reactors. Like other non-weapons uses of HEU, naval reactors present a case where the behavior of the United States may have an appreciable influence over other countries. In recent years, three such countries have expressed serious interest in the development of a nuclear-powered submarine fleet, and two of these nations, India and Brazil, have decided to pursue development. The decision of these two states as to whether to use LEU or HEU could be heavily influenced by the United States' practice. The use of HEU for naval propulsion also has implications for drafting and eventually implementing the terms of a proposed Fissile Material Cutoff Treaty (FMCT).

This chapter highlights the views of the U.S. Navy on conversion prospects and provides lessons learned from the French conversion case. The official position of the U.S. Navy has not changed since 1995, when it reported that conversion "offers no technical advantage to the Navy, provides no significant non-proliferation advantage, and is detrimental from environmental and cost perspectives."[4] This chapter examines the Navy's primary argument against conversion: the economic penalty. While conversion would certainly be costly for the Navy, both the annual costs and the one-time redesigns costs could likely be reduced below the Navy's estimates by making modest changes in the fuel cycle and implementing a prudent schedule for reactor conversion. These alternatives are examined in greater detail below.

All nuclear-powered French naval vessels are currently LEU-powered, despite some contrary assertions in the literature. The French government decided to convert from HEU to LEU fuel for submarines because the continued production of HEU for military applications was uneconomic. The forthcoming class of French submarines represents a substantial improvement over the current class, suggesting that commitment to conversion and sincere research and design efforts can yield high-performing, LEU-powered naval vessels.

Section two of this chapter is a literature review, summarizing the previously available information on naval reactor conversion. Section three explains the methods used to conduct further research on this topic, followed by section four that details the findings. Section five summarizes the lessons learned. Finally, section six offers conclusions and makes recommendations for both policy and future research.

## Literature Review

The most comprehensive study of the proposed conversion of the U.S. fleet was issued in 1995 by the Office of Nuclear Naval Propulsion (ONNP), in response to a Congressional mandate.[5] As noted above, the study concluded that conversion would not provide a significant nonproliferation advantage and would prove costly to the Navy. In order to convert without compromising performance, the ONNP reported that either the reactor would have to be refueled more often, or the core would have to be redesigned to be larger, necessitating a complete ship redesign as well.

Congressional interest in this topic was likely motivated, at least in part, by the revelation in the late 1980s that Brazil planned to construct a secret enrichment plant to support a nuclear naval reactor program. This disclosure raised concerns that if Brazil, a non-nuclear weapons state, were to produce HEU for naval reactors, it would also overcome the main technical hurdle to nuclear weapons. In response to these concerns, the Massachusetts Institute of Technology (MIT) began researching prospects for converting naval reactors to LEU.[6] The fruit of this research, a Masters thesis published in 1990 by Thomas D. Ippolito, represents the only other major, open-source technical assessment of the effects of an LEU core on ship performance.

Ippolito used simulation software to model five reactor cores, two of which ran on LEU (7 percent and 20 percent enrichments), and three of which ran on HEU (97.5 percent enrichment).[7] The fuel used in the model was the French-designed caramel fuel, which France is currently using to power its submarines and sole aircraft carrier at around 7 percent enrichment. His data suggested that the 20%-enriched core could provide a lifetime of 20 years, identical to the HEU core, by expanding its volume to 1.7–2.5 times that of the original HEU core. (By contrast, the ONNP report had cited a need for the LEU core volume to be three times the original.) The Ippolito paper suggests U.S. submarines could easily accommodate this increase in core volume by switching to the French "integral core design." Figure 12.1 illustrates the difference between the French design and the less compact, two-loop design used on U.S. naval vessels.

In an independent assessment, Frank von Hippel performed an alternative LEU design calculation based on the HEU-fueled KLT-40 reactors used in the *Sevmorput* class nuclear Russian icebreaker ships. The KLT-40 reactor, which does not have an integral core, reportedly operates at a power of 135 MWt, which is the approximate power of the reactor in the U.S. Los Angeles-class attack submarines. In his study, von Hippel modified the design to increase the core lifetime to 33 years, commensurate with the desired lifetime for the cores in the newer U.S. Virginia-class attack submarines. The resultant notional dimensions of the LEU reactor core were 50 percent larger than Ippolito's calculation, but still appreciably smaller than the reactor compartment on a U.S. Virginia class submarine. This analysis suggests that if the U.S. Navy pursued a creative, novel reactor design optimized for LEU fuel, it could achieve much more optimistic results than presented in the 1995 ONNP study.[8]

In addition to questions about technical feasibility, the financial consequences of conversion are unclear. The ONNP reported that the conversion would cost an

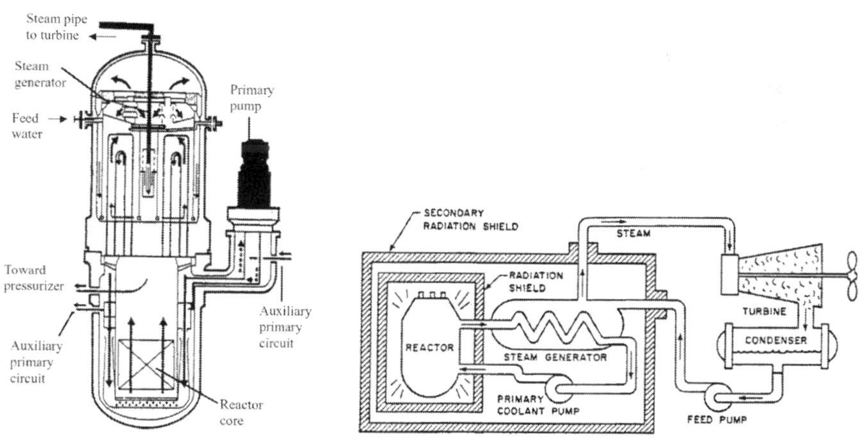


*Figure 12.1* French Integral Core Design (left) vs. U.S. Reactor and Steam Generator Design (right)

Sources: *WINGS Magazine*. Chunyan Ma and Frank von Hippel, "Ending the Production of Highly Enriched Uranium for Naval Reactors," *The Nonproliferation Review* (Spring 2001): 94. U.S. Office of Nuclear Naval Propulsion, *Report on Use of Low Enriched Uranium in Naval Nuclear Propulsion* (Washington DC: 1995): 5.

additional $1.8 billion per year in maintenance costs for the entire fleet (submarines and aircraft carriers), in addition to $4.7 billion in one-time construction costs and an increase of $1.1 billion in annual construction costs. While there would certainly be a high cost associated with conversion, some of this cost might be recouped through reduced security and infrastructure expenditures. The security costs associated with an LEU fuel cycle should decrease relative to those of an HEU fuel cycle, as physical protection standards are commensurate with material attractiveness for weapons use. Although naval vessels still would require high security for other reasons, the security costs for LEU fuel would be lower at the fabrication facility, in transport, and in storage. The ONNP report claims, however, that an increase in spent fuel in the LEU cycle would negate this costs savings. A more rigorous analysis of the real cost of conversion is needed.

The Senate Armed Services Committee 2009 Defense Authorization Report required the Navy to review the prospects of conversion, which indicates that there is renewed interest in the topic.[9] This interest may be spurred by looming negotiations on a Fissile Material Cutoff Treaty (FMCT), which could affect the supply of HEU available for naval propulsion. Von Hippel and Moltz both discuss a "loophole" in the Treaty on Nuclear Nonproliferation (NPT) that establishes "acceptable uses" of HEU for non-weapons states, including naval propulsion.[10] The draft FMCT bans HEU production only for weapons, not for naval propulsion. Nonproliferation advocates fear that unless the FMCT is amended to ban all production of HEU, states could divert HEU from naval propulsion to a weapons program. A Naval Postgraduate School student, in his 2010 thesis, explored the connection between the FMCT and naval propulsion.[11] He reported U.S. Navy concern over possible amendment of the treaty to ban all production of HEU, but concluded that such a modification was highly unlikely.

The existing literature on this topic suffers two weaknesses, namely imbalance and incomplete information. The majority of published work is authored by proponents of naval reactor conversion, with little information from opponents, save the 1995 ONNP report. But the information that the U.S. Navy has provided about the alleged necessity of HEU-powered propulsion is of limited utility due to its lack of transparency. For obvious reasons, the Navy offers few technical details about the operating requirements of its propulsion reactors, making peer review of the claims impossible. Thus a rigorous, independent assessment of the cost-benefit breakdown of naval reactor conversion, including possible consequences for fleet performance, has not been and cannot be conducted with current, publicly available information. My project offers a first step towards such an assessment and recommends additional research and policy analysis.

## Methodology

This chapter builds on the literature to address key questions about the feasibility of, and trade-offs associated with, converting U.S. Navy propulsion reactors from HEU to LEU. The primary question is the validity of the grounds cited by the U.S. military for not converting to LEU. On the technical front, resistance to conversion seems to stem mainly from the expected shorter lifetime of LEU cores. This leads to several subsidiary questions: is it possible to make a lifetime core using LEU, and if so what are the trade-offs entailed, in terms of weight, cost, and other factors? Alternatively, what trade-offs would be entailed by not using lifetime cores in the U.S. Navy's nuclear-powered fleet, including impacts on service time and security risks? On the economic front, what are the projected net savings or costs from using LEU rather than HEU, including from physical security and fuel-cycle expenses? What policy initiatives might facilitate conversion? Finally, what can be learned from the French Navy's experience with conversion of propulsion reactors to LEU fuel?

In order to answer the research questions presented above, I contacted four major categories of interlocutors: American governmental officials, American private sector and NGO representatives, American academics and scientists, and French officials. American governmental officials were contacted at the U.S. Congress, the National Nuclear Security Agency (NNSA) that houses the Office of Naval Reactors (ONR), the office of the Chief of Naval Operations, and national and naval research laboratories. American private sector and NGO contacts included the U.S. Naval Institute and Babcock and Wilcox (one of the Navy's fuel fabricators). Academic and scientific experts contacted were Marvin Miller, Research Affiliate at MIT; Frank von Hippel, professor at Princeton and co-chair of the International Panel on Fissile Material; Armando Travelli, former director of the U.S. government's Reduced Enrichment for Research and Test Reactors (RERTR) program; and Dale Klein, former chairman of the Nuclear Regulatory Commission (NRC). French representatives included Navy and Ministry of Defense officials, as well as scientists at Commissariat à l'Énergie Atomique et aux Énergies Alternatives (CEA) and ANDRA (the national radioactive waste disposal agency). I also contacted two French think tanks and French private sector representatives from Areva, the company responsible for reactor design for the new Barracuda class submarines.

All interviews with French representatives were conducted in person in Paris, where I also met with U.S. Embassy officials from the Office of Defense Cooperation and with the Naval Attaché Adjoint. Bertrand Barré, a scientific advisor to Areva and former CEA employee, was interviewed, as was Mycle Schneider, an energy consultant and member of the International Panel on Fissile Material. Unfortunately Jacques Chenais, the director of nuclear propulsion at CEA, canceled the interview, explaining: "The problem is your state administration [i.e., the Pentagon] thinks that the French are too open on nuclear propulsion. So it's impossible for me to meet with you."

The situation with Mr. Chenais illustrates the major obstacle faced in conducting this research, namely the U.S. Navy's lack of candor and refusal to engage on the issue. NNSA's ONR refused to speak with me, as did scientists at the Y-12 National Security Complex and Knoll's Atomic Power Laboratory, where naval reactors are designed. The scientists themselves at these facilities seemed willing to speak with me, but they were denied permission to do so by ONR. Babcock and Wilcox, the contractor that fabricates fuel for naval reactors, also declined to speak with me. I did manage to establish email contact with Paul Jorgensen, a Senior Operations Manager for the Navy's Advanced Undersea Systems Programs Office.

## Findings

### *Current Fleet Status*

At present the United States has 84 nuclear-powered vessels that run on HEU fuel, including attack submarines (SSNs), ballistic submarines (SSBNs and SSGNs) and aircraft carriers (CVNs). The active nuclear-powered ships in the American fleet are listed in Table 12.1. Maintaining the fleet requires an estimated 2 tons annually of HEU, for which the United States currently relies on a stock of "super-grade" HEU enriched to 97 percent; when that is exhausted, the Navy reportedly will switch to surplus weapons-grade HEU, enriched to 93 percent. A 128-ton stockpile of HEU dedicated to naval propulsion has been established, estimated by an independent analysis as sufficient for "many decades."[12]

Table 12.2 shows information about France's nuclear-powered fleet. All vessels currently run on LEU fuel with an average enrichment of 7 percent. France commissioned its first nuclear ballistic submarine in 1972, *Le Redoutable*, which was probably fueled with LEU, though later *Redoubtable*-class vessels used HEU cores.[13] It is possible that this first French submarine was developed with extensive design assistance from the United States.[14] The last of the *Redoutable*-class subs was decommissioned in 2008, leaving an entirely LEU-powered fleet. Designs of future planned submarines also use LEU fuel.

*Table 12.1* Active U.S. Navy Nuclear-powered Fleet

| | *SUBMARINES* | | | *AIRCRAFT CARRIERS* | | | |
|---|---|---|---|---|---|---|---|
| *Category* | *Class* | *Current* | *Contracted* | *Class* | *Ships* | *Reactors per Ship* | *Reactors per Class* |
| Attack | Virginia<br>Seawolf<br>Los Angeles | 8<br>3<br>44 | 4 | Nimitz<br>Enterprise | 10<br>1 | 2<br>8 | 20<br>8 |
| Ballistic missile | Ohio | 14 | | | | | |
| Guided missile | Ohio | 4 | | | | | |
| Total Subs (one reactor each) | | 73 | 4 | Total | 11 | | 28 |

Note: All propulsion reactors currently use HEU fuel. Four additional Virginia-class submarines are contracted for delivery by 2015.

Source: "US Navy Ships," http://www.navy.mil/navydata/our_ships.asp (accessed March 2011).

*Table 12.2* Active French Navy Nuclear-powered Fleet

| | *SUBMARINES* | | | *AIRCRAFT CARRIERS* | | | |
|---|---|---|---|---|---|---|---|
| *Category* | *Class* | *Current* | *Contracted* | *Class* | *Ships* | *Reactors per Ship* | *Reactors per Class* |
| Attack | Rubis/ Améthyste | 6 | | Charles de Gaulle | 1 | 2 | 2 |
| | Barracuda | | 6 | | | | |
| Ballistic missile | Triomphant | 4 | | | | | |
| | Le Terrible | 1 | | | | | |
| Total Subs (one reactor each) | | 11 | 6 | Total | 1 | | 2 |

Notes: All propulsion reactors now use LEU fuel.

Source: "Equipments, Moyens, Matérial Militaire," 21 December 2010, http://www.defense.gouv.fr/marine/decouverte/equipements-moyens-materiel-militaire (accessed January 2011).

Table 12.3 compares the specifications of a typical American attack sub (Los Angeles class) and a French attack sub. It is clear from the information in the table that the American attack subs are larger than their French counterparts and boast

*Table 12.3* Comparing Published U.S. and French Attack Submarine Specifications

| | *USA - Los Angeles Class* | *France - Rubis Class* |
|---|---|---|
| Enrichment (% $^{235}$U) | 97.3 | 7% |
| Core Endurance (yrs) | 33 | 7 |
| Max Power (MW) | 130 MW / 35,000 shp* | 48 MW/9,500 shp* |
| Max Speed (knots) | 25+ | 25 |
| Power/Displacement ($10^{-2}$ MW/tons) | 1.88 | 1.99 |
| Diving Depth (m) | 300 | 300 |

Note*: shp = Shaft Horse Power

Sources:

Enrichment: Ma and von Hippel, "Ending the Production," 91; High-ranking military official, interview with author, 1 February 2011, Paris, France.

Core Endurance: ONNP, Report on Use, 3; "Nuclear-Powered Ships."

Max Power: High-ranking military official, interview with author, 1 February 2011, Paris, France.

Max Speed: "Attack Submarines-SSN," 10 September 2010, http://www.navy.mil/navydata/fact_display.asp?cid=4100&tid=100&ct=4. "Submarine Proliferation: France Current Capabilities," October 2010, http://www.nti.org/db/submarines/france/index.html (accessed January 2011).

Diving Depth: "SSN-688 Los Angeles Class," http://www.fas.org/programs/ssp/man/uswpns/navy/submarines/ssn688_la.html#spec (accessed March 2011); Ma and von Hippel, "Ending the Production," 91.

more powerful reactors. However, it appears that American submarines actually have a slightly lower power-to-displacement ratio and only a comparable maximum submerged speed and diving depth. It should be noted that the data presented are the published values, so it is likely that the true capabilities of both ships exceed these specifications to an unknown degree.

### *U.S. Navy Resistance to Conversion*

As noted above, the United States possesses a large, highly capable nuclear fleet that runs on HEU fuel. Accordingly, the topic of converting U.S. naval reactors is regarded as unsavory by many U.S. Navy and defense officials. One of the primary goals of this chapter was to determine the specific grounds cited by the Navy for resistance to conversion. The Navy's response, as conveyed by Thomas Dougan of the Naval Reactors Public Affairs office, was that the Navy stood by the position issued in the 1995 report, namely that conversion would be costly and offer no performance advantage to the Navy.

In a phone interview, Marvin Miller explained that the U.S. Navy is resistant to the idea of conversion because it has invested a lot of time and money into the development of a compact, robust reactor that delivers the required power. The requirements for submarine reactor design are extremely stringent, and the Navy has spent the last five decades optimizing its core design with respect to these requirements. In the 1995 report, the Navy cites compactness, crew protection, reliability, ruggedness, maneuverability, endurance, and quietness as essential requirements, and states that failure to meet any one of them would make a core unusable in a ship or would seriously jeopardize the safety of the ship.[15]

Miller notes that in light of these extensive design efforts, the idea of "making changes for the sake of nonproliferation is a joke" to the Navy. He says that the Navy feels that its mission is to support national security, and "national security always trumps nonproliferation."[16] Dale Klein – who has served as both NRC chair and Assistant to the Secretary of Defense for Nuclear, Chemical and Biological Defense Programs – echoed these sentiments in a personal interview. Mr. Klein explained that the Navy has "no philosophical commitment" to the use of HEU but will continue to use the fuel it deems necessary to fulfill its mission.[17]

Of the performance requirements listed above, compactness and endurance are the most often cited as posing obstacles to conversion to LEU fuel. However, a U.S. Naval Institute representative, Paul Jorgensen, noted that maneuverability – the ability of the ship to change power rapidly and frequently – could also be compromised by conversion. He noted that there is a critical difference in the reactor physics of HEU and LEU fuel, which affects a desirable operating characteristic of naval reactors. This characteristic, explained below, is the ability to control the power of the reactor and submarine quickly by varying steam demand without causing an appreciable change in the temperature of the moderator, thereby avoiding the need for human intervention that would delay such control.

The explanation is as follows: in nuclear reactors, power can be controlled either by mechanically moving control rods in the core or by changing steam demand. The latter method means changing the amount of reactor-produced steam that is permitted to enter the turbines to create power for the submarine. For naval propulsion, the latter method is much preferred because it can allow the power to be changed more quickly – without human intervention beyond the request to alter steam demand.

An HEU-fueled reactor can do this, while an LEU-fueled reactor may not be able to. This is due to their respective behavior in response to two changes in temperature: one in the moderator, and the other in the fuel. Because both HEU- and LEU-fueled, pressurized water reactors are water-moderated, the moderator effect is the same for both: namely, an increase in moderator temperature decreases the reactivity, which decreases the power. But the effect of increasing the fuel temperature is different between the two types of reactors, because it depends largely on the U-238 content in the fuel, which is much higher for LEU. As the fuel temperature increases, neutron absorption in U-238 increases appreciably, reducing the reactivity in the core, and thereby reducing the power. For LEU fuel this effect is significant, so increased fuel temperature results in decreased reactivity and power. For HEU fuel this effect is not significant, because of the low U-238 content.

Thus, HEU- and LEU-fueled reactors respond differently to changes in power that are dictated by steam demand (Figure 12.2). In an HEU-fueled reactor, as steam demand is increased ("pulling steam"), the moderator temperature decreases because the steam is carrying heat out of the moderator at a faster rate. This increases the power, due to the moderator effect. But the increase in power in turn increases the moderator temperature, returning it roughly to its original level. Consequently, the initial act of increasing steam demand produces increased power with very little change in moderator temperature.

Conversely, in an LEU-powered reactor, a more complicated feedback occurs, due to the addition of the fuel effect. Initially, as steam demand is increased, the moderator temperature again decreases, resulting in increased power. However, in this case, the fuel temperature effect overwhelms the moderator effect. The increase in reactor power raises the fuel temperature, which reduces the reactivity and thus the reactor power, which further decreases the temperature of the moderator. The net effect is that for an LEU-powered reactor, increasing the steam demand increases the power but decreases the moderator temperature. In order to counter this latter effect, an operator must manually pull a control rod to increase the reactor power and thereby heat the fuel and the surrounding moderator. This retards the rate at which the reactor's power can be changed, because a human action is required, as opposed to merely the physical feedback mechanics in the core that occur much more quickly. The result is a reduction in the maneuverability of the vessel. The need for human intervention also introduces an extra potential source of error.

The feedback loops for both types of reactors are shown in Figure 12.2. $T_0$ denotes the original moderator temperature before the steam demand is increased,

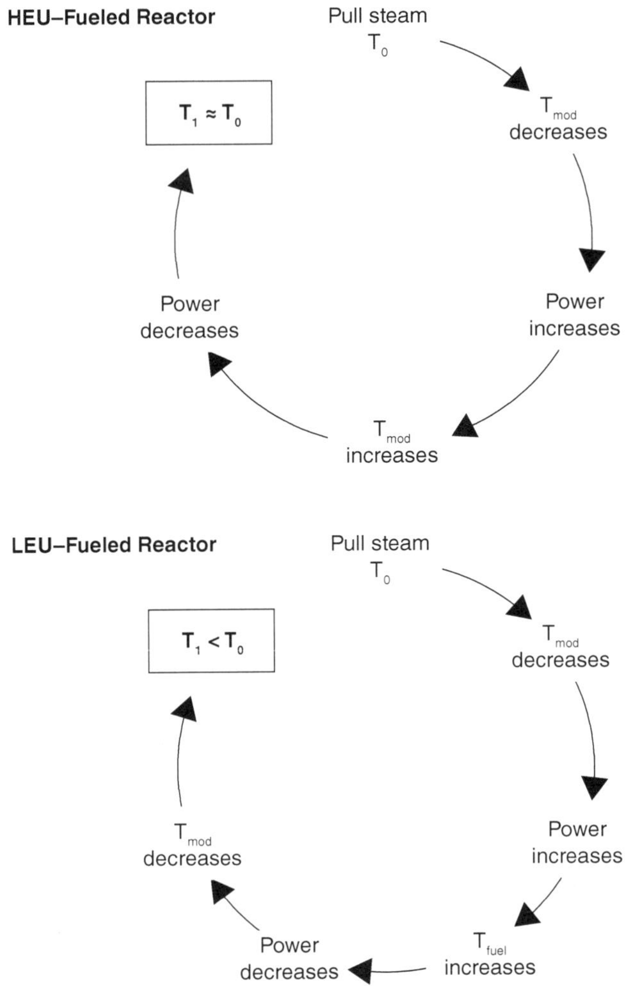


*Figure 12.2* Responses to Increase in Steam Demand by HEU vs. LEU Reactors

and $T_1$ denotes the new moderator temperature after increasing steam demand. $T_{mod}$ and $T_{fuel}$ are used for moderator and fuel temperature, respectively.

Potentially addressing this concern, an academic expert in reactor operations and reactor physics notes that LEU-powered TRIGA research reactors have square-wave operational modes that allow for very fast transitions between power levels (<1 s). While not operational in power reactors, the technology for controlled fast transients in LEU systems appears feasible.[18] Unfortunately, these rapid changes in power level require the use of controls rods and a pneumatic ejection system. Jorgensen suspects that the French use a rod jacking system to produce the needed high-power-demand cycles. He notes that this leads to a complicated control rod system that the U.S. Navy believes is not as inherently easy to operate.

The U.S. Navy prefers not to use automated control rod systems in naval reactors.[19] The reasons for this are likely due to safety and reliability considerations, as they introduce increased possibility of human error and mechanical failure.

While this problem of decreased maneuverability might ultimately have technical solutions, the U.S. Navy has focused on a potentially more stubborn challenge – core endurance – as its primary grounds for opposing conversion. The Navy has invested tremendous resources into the development of lifetime cores for submarines, which are said to "minimize life-cycle cost, minimize demand on support infrastructure, minimize occupational radiation exposure, and maximize ship availability to the fleet for service at sea."[20] Thus, one of the most salient questions for conversion is whether it eliminates the option of a lifetime core, or whether intensive development efforts could produce an LEU lifetime core. Because of the much higher fissile material content in HEU sub fuel, which is 97 percent U-235, the achievable power density is much greater than with LEU fuel, which is less than 20 percent U-235. This allows for production of more power with an HEU core of equal volume. The higher fissile material content of HEU also means the core can continue to produce power for longer periods of time without refueling than an LEU core of comparable size, which has enabled the U.S. Navy to develop lifetime cores. As mentioned earlier, conversion to LEU without significant fuel improvements would require either an increase in core volume or more frequent refueling to compensate for the reduction in fissile material content.

This problem is not unique to naval reactor conversion, as a similar issue was faced and overcome in converting many research reactors to LEU fuel. Research reactor conversion was made possible by the development of higher density fuel, which effectively packs more fissile atoms into the available volume. While the density of most HEU fuel used in research reactors is approximately 1.7 g/cc, the RERTR program has qualified LEU of several times that density and is developing a monolithic Uranium-Molybdenum (U-Mo) LEU fuel of density around 16 g/cc. The latter fuel might appear a good candidate for naval reactors based on the power density it offers, but Marvin Miller explains that the operational constraints of naval reactors preclude its use. Naval reactors operate at temperatures much higher than research and commercial reactors, and the U-Mo fuel exhibits poor metallurgical performance at these temperatures.[21] Miller points out, however, that the Navy has continued to increase the life of the cores, from 20 years to 33 years, with plans to go as high as 45 years.[22] These improvements suggest some increase in density or in core volume has taken place, which indicates that U.S. naval reactors may also have the flexibility to accommodate higher-density LEU cores.[23] Miller suspects that the Navy's reluctance to engage on the issue of conversion may be because research and design of LEU cores is already underway.

### *The French Conversion Case*

While the highest density LEU fuel being designed for research reactors may have limited applicability for naval reactors, the French case indicates that technology transfer from research reactors to naval reactors is possible. The current class of French submarines uses caramel fuel, a higher density fuel originally developed for research reactor conversion. Caramel fuel is a sintered uranium oxide ($UO_2$) fuel clad with zircaloy cladding that has a density of 10.25 g/cc. The active part of the fuel is small $UO_2$ parallelepipeds surrounded by zircaloy. These pieces are welded into a zircaloy plate, and another zircaloy plate is then welded on top.[24] Figure 12.3 shows a top view of a fuel plate before being covered by a second plate.[25] Bertrand Barré, a scientific advisor to Areva and former CEA employee, explained that the welding between individual $UO_2$ sections offers safety advantages, namely the isolation of each fuel "caramel." In addition, this design provides a high cooling surface for the fuel, which affects the speed with which the reactor can change power levels (thus improving the maneuverability of a submarine).[26]

In a personal interview, Barré confirmed that all French nuclear-powered naval vessels currently use low-enriched uranium, which has not been reported consistently in the literature. The cores use caramel fuel and feature three enrichment zones, the average of which is generally reported to be 7 percent for the Rubis submarines. The actual enrichment of each of these zones is classified, as is the actual average enrichment, though all of the fuel used is reportedly LEU. Marvin Miller asserts that the fuel is shuffled every five years and the core replaced every ten years.[27] The ten-year core life referenced by Miller is higher than the seven-year core life referenced by most literature sources. Barré declined to comment on the maintenance cycle of the submarines, citing classification reasons.

The new French attack submarines, the Barracuda class, are under development, with the first ship scheduled to enter into service in 2017.[28] These submarines will be almost twice as large as the Rubis class submarines, reportedly with an

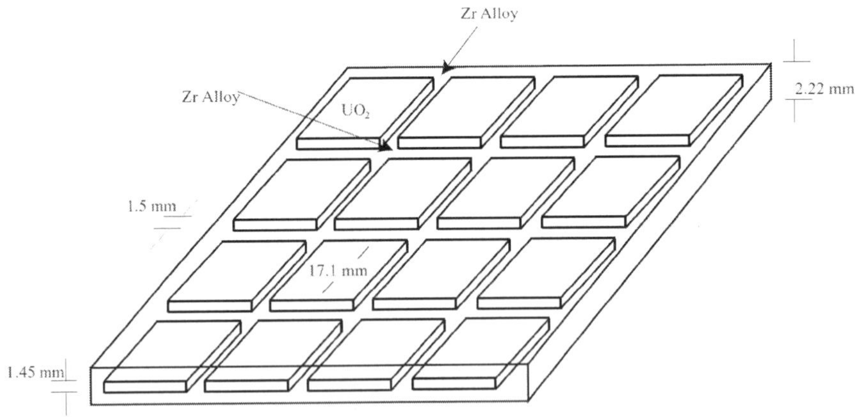


*Figure 12.3* Caramel Fuel Plates Embedded in Zircaloy

Reproduced with permission from Chunyan Ma and Frank von Hippel, "Ending the Production of Highly Enriched Uranium for Naval Reactors," *Nonproliferation Review* 8 (Spring 2001), 97.

increased core life of ten years, compared to the seven years typically reported for the Rubis class. These improvements will be accompanied by a decrease in enrichment to that used for power reactor fuel (< 5% U-235). The submarines will use a modified K-15 reactor, which is the type used in the Le Triomphant class submarines. The fuel will still basically be caramel fuel, with "some improvements," according to Barré.[29]

The K-15 reactors are 150 MW, which represents a three-fold increase over the 48-MW reactors currently used in the Rubis class submarines. As the displacement of the new subs will be twice as large, the power-to-displacement ratio will increase by approximately 50 percent in the new submarines, which should allow for increased core endurance and/or higher speeds. Barré could not comment on the specific performance characteristics of the new submarines. Unconfirmed reports suggest that based on the K-15 reactors, the new submarines should either be able to reach speeds up to 40 knots, or have a core endurance as high as 25 years.[30] Because it is unclear what is meant by a "modified K-15 reactor," it is difficult to predict with any accuracy the specific performance capabilities of the new submarines. It is clear, however, that improvements to core design have enabled increased speed and/or longer core life, again suggesting that sincere efforts by the U.S. Navy to optimize reactor design for an LEU core could yield more impressive results than presented in the 1995 report.

While the Barracuda class submarines will represent a substantial improvement over the Rubis submarines, it must be noted that they still fall dramatically short of U.S. Navy submarine standards. The priorities of the French military differ from those of the United States in this respect. Miller quotes Yves Girard, a member of the French team promoting the sale of the Rubis submarines to Canada, as saying, "You must understand that we don't have an unlimited budget. So our reasonable goal was not to make the best submarine in the world, but to get the best efficiency/cost ratio."[31] By contrast, the goal of the U.S. Navy is, without question, to make the best submarines in the world.

### *Projected Costs and Savings*

Despite compromising ship performance, the French government decided to pursue the development of an LEU-powered naval fleet for economic reasons. All military enrichment in France had been undertaken at a small enrichment plant at Pierrelatte, which was estimated to produce 2–3 tons of enriched uranium over its 29-year operating life (approximately 100 kg/yr).[32] The plant originally had four buildings with different enrichment capacities (2%, 6–8%, 25%, and 90%), and supported both commercial and military applications. Once Eurodif acquired its own capacity to enrich to commercial levels, the Pierrelatte plant closed all but the highest enrichment line and was used only for military applications. This proved uneconomic, so the plant was shutdown, with dismantling commencing in 1998. According to Barré, France no longer has the technical capacity to enrich to levels above 5 percent and does not possess large stockpiles of excess weapons-grade uranium, which forced the initial conversion of submarines to LEU and the plan for commercial-level enrichment in the next generation of propulsion reactors.[33]

In this respect, the French conversion case differs greatly from that of the United States, which has set aside a large stockpile of excess weapons-grade uranium for naval applications. In the United States, there is no economic or fuel-assurance constraint compelling the Navy to convert. In fact, the 1995 U.S. Navy report presented a largely economic argument *against* conversion, citing increased design, fuel-cycle, maintenance and operational costs.[34] While conversion would no doubt prove costly for the U.S. Navy, it seems unlikely that cost concerns are truly driving the Navy's opposition to conversion. A high-ranking U.S. military official confirmed that the military is not driven by cost and that economics is a secondary consideration to performance.[35]

The 1995 Navy report generated two options for conversion: maintaining core volume and reducing core life, or increasing core volume and maintaining lifetime cores. The second option posed greater upfront expenses with a one-time redesign cost of $5.5 billion, but entailed lower annual construction and operation expenses because it placed no additional burden on fleet availability. As a result, over the long run, this total redesign option is preferable economically, as well as from operational, worker-safety and environmental standpoints. Interestingly, when France decided to convert its naval fleet, it chose this option of redesigning "from the ground up" to optimize the reactor's performance with LEU fuel.

If the U.S. Navy were to convert using this option, the 1995 study estimated an annual increase in operating costs of $1.1 billion. But the Navy analysis failed to consider that some of these costs could be recouped through reduced security requirements at several stages of the naval fuel cycle – during fuel fabrication, in transport, and at fuel storage facilities. One of the more staggering costs associated with the naval fuel cycle is the cost of fuel fabrication, which Frank von Hippel estimates at $100,000/lb.[36] Employing the conservative assumption that this price is per pound of heavy metal, fuel fabrication expenses would total $440 million annually. Commercial fuel fabrication costs average around $209/lb of uranium, lower by a factor of 500.[37] It is unclear what expenses are incorporated in the estimated cost of naval fuel fabrication, but the figure may include security costs, which would explain some of the additional expense. The fabrication of LEU should prove less costly due to decreased security requirements, and modest research and design (R&D) efforts should enable technical breakthroughs that reduce this cost further. Even a 50 percent reduction in naval fuel fabrication costs would result in a savings of $220 million annually, which would offset a portion of the additional operating expenses associated with conversion.

Though converting to LEU would certainly offer some savings, such as security costs, the Navy points out that it would also lead to greater waste volume, thereby increasing expenses. At present the U.S. Navy stores all of its spent fuel, primarily at the Naval Reactors Facility at Idaho National Laboratory. The 1995 report notes pressure from Idaho to minimize waste shipments and warns that "tripling to quadrupling the rate of spent Naval fuel shipments also clearly would be objectionable to the state of Idaho."[38] A total of 400 spent-fuel canisters have been generated over the 50-year operating life of the U.S. nuclear fleet, suggesting that the fleet has produced an average of 8 canisters per year.[39] In the 1995 report's ship redesign scenario, annual spent fuel would increase to 23 canisters, exceeding the legal limit of 20 canisters established in the 1995 Idaho Agreement and Consent Order. But this cap might prove flexible, considering that the agreement already has been renegotiated once, in 2008, to extend the deadlines by which naval fuel must be removed from wet and dry storage.[40]

## Summary of Findings

The United States Navy acknowledges that conversion from HEU to LEU fuel for naval propulsion is technically feasible, but asserts that it is uneconomic and would compromise ship performance. The reduction in fissile-material content in the core that accompanies conversion from HEU to LEU decreases the achievable power density of the core, posing technical obstacles that must be overcome if current submarine performance is to be maintained. Specifically, the greatest technical concerns are over the possibility of creating a small, lifetime core with LEU that still allows for rapid changes in power. Most of these same technical obstacles have been faced in converting research reactors from HEU to LEU, but have been largely overcome by the development of higher density LEU fuel, which allows more fissile atoms to be packed into a certain core volume. Unfortunately, the highest density fuel, monolithic U-Mo, is not a suitable candidate for use in naval propulsion, due to the operational constraints of naval reactors.

While U-Mo may not be practical for use in naval reactors, the French have developed a uranium oxide LEU fuel with density about six times greater than the formerly standard uranium oxide HEU fuel used in research reactors.[41] This fuel is currently used in all French naval vessels, providing proof of concept for the viability of naval reactor conversion. When they converted from HEU, the French chose to design new naval reactors from the ground up to optimize the design for use of LEU fuel, rather than modifying an already existing design. Though the French have successfully developed reliable and capable submarines with LEU fuel, their experience underscores the U.S. Navy's concern that conversion would compromise either core life or core volume. In the case of the French conversion, core life was sacrificed, producing submarines that reportedly need to be refueled every 7–10 years.

The ultimate impetus for the French conversion to LEU powered submarines was economic. The French enrichment plant for HEU was too costly to justify continued operation, so the French government shut it and converted the propulsion reactors to an average of 7%-enriched LEU fuel. France no longer possesses the technical capacity to enrich to levels above 5 percent and has no stockpile of HEU for naval purposes.

Due to this enrichment constraint, the next generation French submarines, which are currently under development, will reportedly use modified K-15 reactors running on fuel enriched to commercial levels below 5 percent. Assuming the modified K-15 reactor maintains a rating of 150 MW, the new larger sub will still have a higher power-to-displacement ratio than the Rubis class submarines. This would yield a longer core life of 10 years, up from the current 7 years, despite a roughly 3 percent drop in fuel enrichment. Such substantial improvements are promising. If France's cost-constrained R&D can produce a 43 percent increase in core life despite an increase in ship displacement and decrease in enrichment, then sincere American research efforts probably could develop an excellent submarine with 20%-enriched fuel.

## Conclusions and Recommendations

The priority of the United States Navy is developing and maintaining the best submarines in the world. Accordingly, the advocacy of converting naval reactors from HEU to LEU will prove fruitless if conversion would substantially compromise ship performance. Developing LEU naval reactors that perform to the level of the current HEU-fueled reactors would require major research, development, and design efforts by the U.S. Navy. That is not to say that such conversion is impossible, but at present the Navy has no obvious incentive to pursue such conversion. Ample resources have already been invested to obtain the current reactors, which meet all of the Navy's stringent performance requirements, and to assure its fuel supply.

U.S. naval propulsion thus represents a major barrier to the global phase-out of HEU, at least in the near-term. This could pose several problems in the international arena. First, Miller notes that continued use of HEU by the U.S. for naval propulsion legitimizes general use of HEU internationally, thus undermining worldwide conversion efforts. More immediately and specifically, it raises nationalism issues among emerging naval nuclear powers. Both India and Brazil are said to be developing nuclear submarine fleets, and both initially wanted to use HEU to power these fleets, prompting U.S. opposition. Miller explains that on issues like nuclear propulsion, the United States and several other nuclear-weapons states have been "advocating water and drinking wine." Neither India nor Brazil is insensitive to proliferation concerns, Miller says, but neither wants to face a double standard, especially with respect to military operations.[42] Thus the continued use of HEU for naval propulsion by the United States undermines the international norm against non-weapons uses of HEU.

The use of HEU for naval propulsion also poses problems for the drafting and implementation of a Fissile Material Cutoff Treaty. If the treaty permits the production of fissile material for naval propulsion, there will be a loophole for states to continue producing HEU, ostensibly for naval propulsion, but which could be diverted to weapons. At the least, such terms would greatly complicate the verification and enforcement of a FMCT.

### *Recommendations*

As the information above makes clear, failure of the U.S. Navy to convert its propulsion reactors from HEU to LEU fuel has potentially large consequences for the international arena. Of course, so could any compromise in submarine performance. Accordingly, the policy recommendations with respect to potential U.S. naval reactor conversion are as follows:

1. No effort should be taken to convert the current submarine fleet.
2. The timeline for any conversion should be several decades (~40 years), but consideration of the issue should start soon.
3. A rigorous analysis of the value and feasibility of conversion should be conducted by an independent third-party with access to classified information.
4. If conversion is deemed feasible and prudent, economic incentives should be provided to the Navy to help alleviate the financial burden.

As mentioned previously, the Navy's HEU stockpile is large enough to sustain the current fleet for "many decades." The Navy should continue to use this stockpile in the near term. Submarines that are currently in service or under development have been designed and optimized to use HEU fuel. Efforts to convert these submarines would require major redesign efforts, retrofits, and overhauls, which would be costly at best, and potentially unsafe at worst. The high conversion costs estimated in the 1995 report seem to reflect this scenario, based on adapting current designs to LEU use. Instead, new LEU-powered submarines should be designed from the ground up, as was the case with the French Barracuda class. This effort should be integrated with any current or future Navy plans to design new submarines. Research and development efforts for propulsion-reactor conversion should begin now, so that after two or three design cycles, when the Navy is ready to finalize a new class of submarines, the conversion work will be advanced enough to be incorporated.[43]

Before any R&D work is done, however, a more transparent analysis of conversion should be conducted. As Miller said of the Navy's position on this issue, "Their contention … is both strongly held and impossible to verify without access to classified information."[44] For an assessment to generate truly meaningful results, it must be conducted by an independent third party, given access to classified information. Further, the assessment should rigorously consider the costs and benefits from both economic and security perspectives.

If a rigorous assessment indicates that conversion is potentially feasible, beneficial, and affordable, research and development should be undertaken to develop the necessary technology to enable conversion. Additionally, research on ways to reduce naval fuel-cycle costs, especially fuel fabrication costs, should be conducted. A small investment in fuel fabrication research could result in major costs savings. Finally, all of these efforts must be undertaken as collaborative efforts between the Navy, the government and the national laboratories, as substantial innovation will be required to reduce fuel enrichment by 75 percent without compromising ship performance.

The R&D required to make conversion possible would be costly, as would the conversion process itself, which serves as a major disincentive for the U.S. Navy. To combat this problem, funding should be provided to the Navy to help alleviate additional expenses associated with both R&D costs and increased operational expenses if conversion is implemented. Such spending would be justified to support a global HEU phase-out that could significantly reduce risks of nuclear proliferation and nuclear terrorism.

## Notes

1 "Nuclear-Powered Ships," last modified 19 November 2010, http://www.world-nuclear.org/info/inf34.html (accessed on 27 September 2012).
2 Chunyan Ma and Frank von Hippel, "Ending the Production of Highly Enriched Uranium for Naval Reactors," *The Nonproliferation Review* (Spring 2001): 92, accessed October 2010.
3 Ole Reistad and Styrkaar Hustveit, "HEU Fuel Cycle Inventories and Progress on Global Minimization," *The Nonproliferation Review* 15 (2008): 271, accessed November 2010.
4 United States Office of Nuclear Naval Propulsion, *Report on Use of Low Enriched Uranium in Naval Nuclear Propulsion,* 1995, 104th Cong.:1.
5 ONNP, *Report on Use,* 1–40.
6 Marvin Miller, "The Use of HEU in Naval Nuclear Reactors and Its Implications for a Fissile Material Cutoff Treaty (FMCT)" talk given in Geneva, Switzerland, 26 September 2003.
7 Thomas D. Ippolito, "Effect of Variation of Uranium Enrichment on Nuclear Submarine Reactor Design" (M.S. thesis, Massachusetts Institute of Technology, 1990).
8 Ma and von Hippel, "Ending the Production," 96.
9 Senate Committee on Armed Services, National Defense Authorization Act for Fiscal Year 2009, S. Rep. No. 110–355, at 515 (2008).
10 Ma and von Hippel, "Ending the Production," 86; James Clay Moltz, "Closing the NPT Loophole on Exports of Naval Propulsion Reactors," *The Nonproliferation Review* 6 (1998): 108, (accessed November 2010).
11 Marion Burgess Jr., "The Proposed Fissile Material Cutoff Treaty (FMCT) and Its Potential Impact on U.S. Navy Nuclear Propulsion Programs" (M.S. thesis, Naval Postgraduate School, 2010).
12 Ma and von Hippel, "Ending the Production," 92.
13 "Histoire de la Marine Nationale," last modified 15 December 2010, http://www.defense.gouv.fr/marine/decouverte/patrimoine/histoire/histoire-de-la-marine-nationale/histoire-de-la-marine-nationale (accessed on 27 September 2012); "Fabrication of Uranium-based Fuel," last modified 6 January 2007, http://www.francenuc.org/en_chn/fabricationu_e.htm (accessed on 27 September 2012).
14 High-ranking military official, interview with author, 1 February 2011, Paris, France.
15 ONNP, *Report on Use*, 3.
16 Marvin Miller (Research Affiliate, MIT), interview with the author, 1 December 2011.
17 Dale Klein (former NRC Chair; Faculty, University of Texas at Austin), interview with author, 28 March 2011.
18 Academic reactor operations expert, email message to author, 19 April 2011.
19 Jorgenson, email.
20 ONNP, *Report on Use*, 3.
21 Miller, "The Use of HEU."
22 ONNP, *Report on Use*, 9.
23 Miller, "The Use of HEU."

24 J.P. Schwartz, "Uranium Dioxide Caramel Fuel: An Alternative Fuel Cycle for Research and Test Reactors," paper presented at International Conference on Nuclear Non-proliferation and Safeguards, New York, 22–25 October 1978.
25 Ma and von Hippel, "Ending the Production," 97.
26 Bertrand Barré (Scientific Advisor, Areva), interview with author, 2 February 2011, Paris, France.
27 Miller, interview.
28 "SSN Barracuda Nuclear-Powered Attack Submarine, France," accessed March 2011, http://www.naval-technology.com/projects/barracuda/ (accessed on 27 September 2012).
29 Barré, interview.
30 "Submarine Discussion Board," last modified 18 January 2005, http://www.strategypage.com/militaryforums/462-1488/page1.aspx (accessed on 27 September 2012).
31 Miller, "The Use of HEU."
32 "Rhone-Alpes," accessed 28 May 2009, http://www.francenuc.org/en_sites/rhone_tri_e3.htm (accessed on 27 September 2012).
33 Barré, interview.
34 ONNP, *Report on Use*, 1.
35 High-ranking military official, interview.
36 Frank von Hippel, "Global Cleanout of Highly Enriched Uranium: the Challenges," talk given at LBJ School, University of Texas, Austin, TX, October 2010.
37 "Nuclear Fuel Cost Calculator," last modified 23 November 2009, http://www.wise-uranium.org/nfcc.html (accessed on 27 September 2012).
38 ONNP, *Report on Use*, 16.
39 "Naval Reactors: United States Naval Nuclear Propulsion Program," talk given at the first public meeting of the Blue Ribbon Commission, Washington D.C., 25–26 March 2010.
40 "Naval Reactors".
41 Most standard HEU-powered research reactors used fuel with uranium density around 1.7 g/cc. There is a good chance, however, that American submarines are already using fuel more dense than this.
42 Miller, interview.
43 Pavel Podvig, "U.S. Navy plans for a new class of ballistic-missile submarines," 24 January 2010, http://www.fissilematerials.org/blog/2010/01/us_navy_plans_for_new_cla.html (accessed on 27 September 2012). Two years later, the Pentagon reportedly called for "delaying procurement of a new, proposed Ohio-class replacement nuclear-armed submarine by two years." See, Daryl G. Kimball, "Pentagon Budget Delays Significant Decisions on Trimming Excessive, Expensive Cold War Nuclear Forces," 26 January 2012, http://armscontrolnow.org/2012/01/26/pentagon-budget-delays-significant-decisions-on-trimming-excessive-expensive-cold-war-nuclear-forces/ (accessed on 27 September 2012). Nevertheless, the FY 2013 budget request reportedly contains $1 billion for design of the propulsion reactor for the new submarine, in addition to $565 million for design of the submarine (on top of $2 billion spent so far). See, Walter Pincus, "Nuclear weapons modernization: not fast enough for Kyl," *Washington Post*, 27 February 2012.
44 Miller, "The Use of HEU."

# 13 Russia

## Naval Propulsion

*Yaroslav Primachenko*

This chapter addresses the technical, political, and economic prospects – and obstacles – for conversion of Russia's military naval propulsion reactors from highly enriched uranium (HEU) to low enriched uranium (LEU) fuel. This is a major global security issue because Russia's nuclear Navy requires each year an estimated 570 kilograms of fresh HEU – roughly one-seventh of total worldwide consumption and sufficient for many nuclear weapons if stolen or diverted.[1] The potential elimination of so much HEU from global commerce would significantly reduce risks of nuclear proliferation and nuclear terrorism. Moreover, it would set a very strong precedent for a total global phase-out of non-weapons HEU, because Russia's navy poses some of the greatest political and economic challenges for converting reactors from HEU to LEU, as detailed in this chapter.

Amongst the new findings of my research is that Russia actually plans to *increase* the enrichment of the HEU in its naval reactors, contrary to the objectives of a global HEU phase-out. Its naval reactors currently use HEU of 20 percent to 70 percent enrichment, but Russia plans to increase the enrichment to extend the life of the reactor cores so that their refueling coincides with the vessels' scheduled major maintenance.

Currently, Russia has no plans to convert existing naval reactors to LEU because it perceives no incentive to meddle with a proven design and fuel cycle. Likewise, there are no plans to use LEU fuels in new vessels because there is a concern that the use of such fuel would affect vessel performance – a vital consideration in naval propulsion. Moreover, the conversion of any naval reactors to LEU fuel would necessitate potentially costly modification of the RT-1 reprocessing plant at Mayak, which is designed to process HEU spent fuel.

My findings also contradict past reports that attribute Russia's reluctance to convert naval reactors to two factors: 1) the inability to produce substitute LEU fuels; or 2) concern that conversion of naval reactors would exacerbate a backlog of decommissioned vessels and reactors awaiting dismantlement. My research indicates that Russia's Electrostal facility would be able to manufacture viable LEU fuels if requested. Moreover, only about 10 to 15 decommissioned vessels are currently in the queue for dismantlement – all scheduled to be dismantled by 2015 – so this is not a substantial deterrent to conversion.

Russia's officials contend that the security of their naval HEU has improved significantly since the first years after the Cold War, and the facts seem to support them. Contrary to the turbulent Yeltsin era, there is no evidence of theft in recent years. This may reflect the success of the U.S. "Material Protection, Control & Accounting" (MPC&A) program, which has supported upgrades in Russia. Despite such improved security, however, the best protection against theft of naval HEU would be converting Russia's fleet to LEU fuel, which the United States could promote through additional actions and assistance.

The next section of this chapter reviews the existing literature on the topic. The following sections outline the research methodology, present detailed findings, and then summarize the highlights. The chapter concludes with policy recommendations.

## Literature Review

Substantial information has been published on past and present Soviet and Russian military naval nuclear propulsion. In 2002, Maerli reported the supposedly higher enrichment level (90 perrcent) of Russia's newer submarine models and the falling demand for naval fuel due to Russia's economic situation.[2] In 2005, Reistad, et al., traced the Soviet/Russian nuclear fleet from its inception to that year, including the reactors and possible fuels used.[3] While that report contains a lot of speculation, as do many documents on this topic, it is one of the best available. A subsequent article, "HEU Fuel Cycle Inventories and Progress on Global Minimization," by Reistad and Hustveit, describes the throughput of HEU in Russian naval propulsion reactors from 1965 all the way through projected use in 2020, and lists core inventories for HEU-fueled vessels.[4]

In light of economic pressures, the Russian Navy is focused on maintaining only its most modern assets. A large contingent of submarines has been decommissioned due to lack of funds to keep them in service. Accordingly, the remaining fleet is expected to endure longer deployments, which will decrease its service life.

The Bellona Foundation is an international environmental NGO that is a major source of information about Russia's nuclear Northern Fleet, which patrols near the Arctic Ocean. In 2001, the foundation issued a report that examines in great detail the decommissioning of Russia's nuclear submarines. The report addresses the management of spent nuclear fuel including storage bases, transportation to the Mayak reprocessing facility, and concerns related to security, storage, and transportation.[5]

Pavel Podvig claims that about 100 tons of HEU is allocated to the Russian Navy, but only the HEU that has been converted to naval fuel is transferred to the Navy. The rest is kept by Rosatom. Naval fuel is manufactured only at one facility in Russia, the Machine-building Plant (MSZ) in Electrostal – a city just outside Moscow.[6] In its 2010 report, the International Panel on Fissile Materials confirms that spent fuel is reprocessed at the Mayak facility.[7]

Some instances of theft have been reported at Russian naval shipyards.[8] However, a DOE-Russia cooperative MPC&A program has helped Russia secure, or at least bolster physical protection at, most naval sites.[9]

Nearly all Russian military nuclear propulsion reactors that are still in service are reported to use HEU enriched to an average of 21 percent to 45 percent, with one other design using higher enrichment. Older nuclear submarines are powered by four reactors – VM-4S, VM-4SG, VM-4A, and VM-5 – with average HEU fuel enrichment of 21–45 percent. The new *Borei* class submarine is thought to have a single OK-650 nuclear reactor, also with fuel enrichment estimated between 21–45 percent. The single *Kirov* missile cruiser is thought to have two reactors with HEU fuel enrichment of 55–90 percent. An average submarine core contains about 115 kg of U-235 and has a life expectancy of eight years.

Some experts speculate that Russia's existing submarine reactors could be converted to LEU fuel because Russia has already designed a related reactor, the KLT-40s, to use LEU fuel in a floating nuclear power plant. It has been estimated that an LEU-fueled, Russian submarine reactor could yield a core life of 7 to 11 years.[10]

A 2007 Russian article provides insight into the government's priorities for design of future reactors for nuclear submarines.[11] Next-generation propulsion reactors must have a single design, increased reliability (meaning decreased power density in the core to reduce wear), an isolated reactor compartment capable of automatically stopping and cooling the reactor, and spring-activated, neutron-absorbing control rods. Furthermore, the plan is to decrease the amount of liquid and solid radioactive waste, and to develop a core with 1.5 to 2 times the energy reserves of present ones, ideally enabling a core that lasts for the lifetime of the vessel. There is also interest in making components as interchangeable as possible between civilian icebreakers and naval submarines. This would reduce the cost of producing new cores, which is an important factor in light of Russia's economic situation.

Finally, OKBM – Russia's main design and production facility for propulsion reactors – has a prototype reactor (RITM-200) for the new generation of icebreakers. This reactor uses LEU fuel and is nearly 50 percent smaller and lighter than the KLT-40 reactor now used in icebreakers. The RITM-200 is reportedly very versatile and can be used on a variety of platforms. It may also be the first Russian reactor to have the steam generator inside the reactor vessel – a space-saving, "integral" design – as France does in the reactor for its *Rubis*-class submarines. If this new LEU-fueled, Russian icebreaker reactor proves to be less expensive to produce and maintain, as advertised, then it might also facilitate conversion of Russian nuclear submarines to LEU fuel.[12]

The strengths of the existing literature are in describing the historical development of the Russian nuclear military fleet, the reactors used in the past, and the reactors that may be powering current vessels. It also superbly covers the Russian Navy's HEU throughput, including future projections, and the environmental legacy. But the literature falls short both in explaining the past – why Russia designed its vessels and reactors as it did – and in forecasting the future of Russian naval propulsion. Moreover, it fails to address Russia's potential interest in converting existing vessels, or designing future ones, to run on LEU fuel. There is no published analysis of the factors that might encourage Russia to convert its Navy to LEU fuel, including ways the United States could promote such conversion, nor of the likely obstacles. This chapter attempts to investigate these and other unknowns, assess the assumptions of the existing literature, and lay the groundwork for further research.

## Methodology

My primary research questions for this chapter were as follows:

- Why has Russia chosen to use significantly lower enrichment HEU fuel in its military naval vessels than have the United States and United Kingdom, and would this facilitate conversion to LEU?
- Does Russia plan to convert its existing naval reactors from HEU to LEU fuel? Has it ever considered doing so? If it does not currently plan to do so, could any factors – domestic or foreign – incentivize conversion of existing reactors? What are the technical, economic, political, or other obstacles to such conversion?
- Does Russia plan to design its next generation of naval reactors to use LEU fuel? If so, why? If not, why not, and could any factors incentivize it to do so?
- How could the United States promote conversion of Russia's existing or future naval propulsion reactors from HEU to LEU?
- Is Russia's backlog of spent fuel and reactors from decommissioned submarines a significant deterrent to converting existing naval reactors to LEU because doing so would add to the backlog? What is the status and progress of Russia's efforts to dismantle already decommissioned submarines?
- Does Russia really plan to use LEU fuel in the reported RITM-200 reactor for the next generation of civilian icebreaking ships and, if so, could a variant of this reactor also propel submarines with LEU fuel?

Addressing these questions required going beyond the existing literature, so I tried to interview experts who had worked on or studied Russian nuclear propulsion. This was challenging because many Russians with potentially insightful information were not at liberty to speak regarding a topic that is so highly sensitive. But in Moscow I was able to interview Dr. Alexander Dmitriev, an engineer at the Scientific & Engineering Center on Nuclear and Radiation Safety. In addition, I conducted phone interviews with Aleksandr Nikitin, a Bellona NGO associate, and Pavel Podvig, an affiliate at Stanford University's Center for International Security and Cooperation. Furthermore, I was able to acquire information from Vladimir Kuznetsov, who currently works at the IAEA's Nuclear Power Division.

## Findings

*Why has Russia chosen to use significantly lower enrichment HEU fuel in its military naval vessels than have the United States and United Kingdom, and would this facilitate conversion to LEU?*

Russia's decision to use lower enriched fuel in its naval vessels stems from a combination of many factors. In some respects, it is a matter of technological path dependency. During the first few years of the Soviet Union's development of its nuclear fleet, the main concern was keeping up with the United States. The goal was to produce as many as possible, relatively reliable, nuclear vessels – without delaying for testing and sea trials that might optimize them. Because of concerns regarding reactor reliability, it was decided that keeping enrichment at a relatively low level would be safer for the core. Moreover, at the time the Soviets deemed it technically impossible to produce a lifetime core even with very high enrichment HEU, so the main consequence of lower enrichment was merely to require somewhat more frequent changes of the core.[13]

Much of the existing literature incorrectly characterizes the enrichment level of the reactor fuel in Russian military vessels. The literature typically assumes that the average enrichment level in the reactor is uniformly distributed throughout the core. To the contrary, Russian naval reactor cores consist of several zones of varying enrichment. In contemporary vessels, enrichment in some parts of the core may reach as high as 70 percent. So, while the average enrichment in the core may be approximately 40 percent, it is not entirely true that Russian military vessels run on dramatically lower enriched fuel than their U.S. and UK counterparts, which use super-grade HEU enriched to 97 percent. Although the average fuel enrichment in Russian naval vessels is indeed closer to LEU, it would be wrong to assume that the enrichment is so low throughout the core that conversion to LEU would be a trivial challenge.[14]

*Does Russia plan to convert its existing naval reactors from HEU to LEU fuel? Has it ever considered doing so? If it does not currently plan to do so, could any factors – domestic or foreign – incentivize conversion of existing reactors? What are the technical, economic, political, or other obstacles to such conversion?*

Russia has no plans to convert its existing naval reactors, nor has it ever considered doing so. The Russian Navy has spent many years and much money improving its reactor design, and has never perceived or considered any political constraint on the enrichment level of its fuel. Russia has all the HEU it needs, and it does not see a proliferation risk in using HEU fuels, as elaborated below. Russia feels it is using a proven reactor design with a proven fuel cycle, and any deviations from this could introduce costs and uncertainty regarding reliability and performance. In the absence of any initiative by the United States to convert its own naval fleet to LEU fuel, the Russian Navy perceives no pressure or incentive to convert its reactors.[15] Thus, the biggest obstacle to converting Russia's existing naval reactors to LEU fuel is Russia's lack of perceived motivation or incentive to do so, which makes moot other potential hurdles – whether technical, economic, political, or otherwise. If some incentive for conversion were introduced, then feasibility studies could be conducted to evaluate these other potential obstacles and means for overcoming them.

*Does Russia plan to design its next generation of naval reactors to use LEU fuel? If so, why? If not, why not, and could any factors incentivize it to do so?*

As with the existing fleet, Russia does not appear even to have considered using LEU fuels in future military naval reactors. None of the Russian experts I interviewed considered this proposal serious or worthy of exploration. Dr. Dmitriev claimed that none of the existing LEU fuels, nor even the higher-density fuels under development, could give the vessels the same performance characteristics as HEU fuel.[16]

To the contrary, Russia is looking to *increase* the enrichment of HEU fuel in its next generation of naval reactors. The main reason is to extend the life of each core so that it coincides with the schedule for major maintenance of its vessel (the exact period for which is not publicly known), thereby maximizing efficiency. Having the ability to refuel the vessel while performing other routine maintenance would significantly reduce costs, decrease time in port, and lengthen the deployment period. Russia thus sees an increase in fuel enrichment as an economical way to extend deployment of its naval vessels, helping compensate for the country's transition to a smaller nuclear naval fleet.[17] As with its existing fleet, Russia does not perceive any incentive to use LEU fuel in its next generation of naval reactors. As further evidence, Podvig says that all of Russia's fuel-design laboratories rejected the offer of funding from the U.S.-based, non-profit organization, the Nuclear Threat Initiative, to help develop LEU fuels.[18]

*How could the United States promote conversion of Russia's existing or future naval propulsion reactors from HEU to LEU?*

So long as the United States has more nuclear-powered naval vessels, fuels them with HEU of higher enrichment, and has no plans to convert to LEU fuel, there is little prospect of persuading Russia to convert its naval fleet to LEU fuel. But even if the U.S. Navy were to convert its reactors to LEU, Podvig says Russia might not automatically follow suit. In light of Russia's large stockpile of HEU, prioritization of vessel performance, and relative lack of concern about nuclear security, additional incentives would probably be required.[19]

Podvig sees three avenues by which the conversion to LEU fuel of Russian military naval reactors fuel might be facilitated: 1) If the propulsion reactors in Russia's closely related civilian icebreaker fleet were converted to LEU; 2) If Russia succeeded in commercializing its closely related, floating nuclear power plants, possibly requiring U.S. funding; or 3) If there were a U.S.-assisted, joint research program to build a new generation of naval reactors using LEU fuel in both countries. Any of these actions would help offset Russia's cost of developing a new, LEU-fueled naval reactor, and the required fuel. The last action would also signal that the United States is committed to a universal phase-out of non-weapons HEU, including for naval propulsion, and not looking to exempt itself.[20]

But Russian experts doubt that the United States is sufficiently concerned about the Russian Navy's continued use of HEU fuel to commit resources to phase it out. Nikitin asked, "What is the incentive for the United States to have the Russian

Navy run on LEU?" He says that the situation is much safer now than it was during the turbulent Yeltsin years. The last time he heard about any material being stolen was more than a decade ago, and he claims that the security for naval reactors and fuels is on par with the security for nuclear weapons. Furthermore, he argues, there are fewer vessels now, which means less fuel and refueling, which translates into less material vulnerable to falling into the wrong hands.[21]

Dmitriev agrees that the Yeltsin years were the most dangerous time for nuclear proliferation, which he says is now a non-issue. He adds that even back then, thefts were committed by amateurs who stole because nuclear material was viewed as something valuable. Most of the time, these thieves did not know what they had, what it could be used for, or where to find a buyer. In fact, finding a buyer was the most difficult aspect for thieves of nuclear material.[22]

Russian officials also claim that the security threat from their Navy's use of HEU fuel is further reduced because the spent fuel is proliferation-resistant. Their rationale is that the spent fuel contains only minute concentrations of plutonium, so thieves would have to reprocess a large amount of spent fuel to extract sufficient plutonium for a weapon, which is unrealistic for non-state actors.[23] While technically true for extracting plutonium, this ignores the prospect of reprocessing a smaller amount of fuel to extract a sufficient amount of HEU for a nuclear weapon. More importantly, it ignores the even greater risk from fresh HEU fuel. In addition, both of these latter risks would increase if Russia were to proceed with its reported plan to increase the enrichment of HEU in its next generation of naval reactors.

Finally, and ironically, the U.S.-funded MPC&A program may have given Russia a false sense of security. Dmitriev claims that there is virtually no proliferation or terrorism risk at Russia's nuclear facilities because the "Americans came in here and spent a lot of money on security."[24] To be sure, during the Yeltsin years, the MPC&A program greatly improved the security of Russia's nuclear materials. But it is impossible to know if these improvements are adequate. Moreover, technological security mechanisms such as alarms, cameras, and locks can be – and sometimes are – simply turned off.

*Is Russia's backlog of spent fuel and reactors from decommissioned submarines a significant deterrent to converting existing naval reactors to LEU because doing so would add to the backlog? What is the status and progress of Russia's efforts to dismantle already decommissioned submarines?*

The literature indicated that one reason for Russia's reluctance to convert its existing nuclear naval fleet to LEU fuel was an alleged backlog of already decommissioned Soviet-era submarines awaiting dismantlement, including their reactor compartments. But my primary research did not confirm this hypothesis. Nikitin says that only about 10 to 15 retired submarines are in the dismantlement queue.[25] Podvig further explains that these decommissioned vessels are already in various stages of dismantlement, which could be completed as early as 2015.[26]

Reprocessing the spent fuel from the dismantled reactors will take somewhat longer, but there is no backlog of such spent fuel awaiting reprocessing. To the contrary, Podvig reports that the RT-1 reprocessing plant at Mayak has surplus capacity to process spent HEU. In fact, this under-utilization of its reprocessing line could motivate Mayak to oppose any Russian conversion of naval reactors to LEU fuel because, without spent HEU fuel to reprocess from Russian naval reactors, Mayak could lose a major source of revenue. Mayak could sustain this revenue only if the Navy decided to reprocess the LEU spent fuel and if Mayak added or converted a line capable of processing it. (The prospects of foreign spent fuel being shipped to Mayak for reprocessing, which could compensate to some degree for any revenue loss, are uncertain.)[27]

*Does Russia really plan to use LEU fuel in the reported RITM-200 reactor for the next generation of civilian icebreaking ships and, if so, could a variant of this reactor also propel submarines with LEU fuel?*

Based on my field research it appears likely that a version of the RITM-200 will also power the next generation of Russian military nuclear vessels. Indeed, a precursor of this reactor may already be powering the new *Borei* class submarines.[28] However, even though the RITM-200 is slated to use LEU fuel in icebreakers, the reactor is expected to use use higher enriched fuel in military vessels. The RITM-200 is a versatile reactor that – like some previous Russian naval reactors – will utilize cassette-type fuel, which can accommodate varying enrichment levels, including in active zones of the core particular to specific applications.[29]

Accordingly, the current prospects for converting Russia's naval reactors to LEU fuel appear dim. Technically, it probably can be done, in light of the reported use of LEU fuel in Russia's prototype KLT-40S floating reactor,[30] and a Russian feasibility study on converting its KLT-40 icebreaker reactor to LEU fuel.[31] However, it is difficult to assess confidently the feasibility of conversion when the technical specifications and characteristics of the reactors in question are secret. Unless a revolutionary new LEU fuel were developed, conversion to LEU fuel would reduce the operational characteristics of Russia's vessels, including notably the core life of the next generation naval reactor, which is a priority for Russia's Navy.

For existing Russian naval reactors, one theoretical option would be conversion to caramel-type LEU fuel, like that used in France's *Rubis* class submarines. But, again, it is difficult to evaluate the feasibility without more reliable information about Russia's reactors. Furthermore, Russian officials generally view *Rubis* submarines as inferior. (Ironically, some Russian officials also doubt that the French submarines run on LEU fuel, although that is well documented in the preceding chapter of this volume.)[32]

## Summary of Findings

The use of medium-enriched, rather than 90%+ enriched, HEU fuel in the cores of Russian military nuclear vessels was driven by an early Soviet decision to maximize reliability and safety of its naval reactors, and has been sustained by path dependency. The reactor cores contain various active zones where the enrichment of the fuel can vary, with the highest enrichment being around 70 percent.

Presently, Russia has no plans to convert its existing military naval reactors to LEU. The main reasons for this are the preference for sustaining proven and reliable technologies, the lack of perceived proliferation concerns regarding the use of HEU, the lack of incentives and pressures to change the status quo, and the lack of plans for conversion by the U.S. Navy.

Likewise, Russia does not plan to use LEU fuels in its future military naval reactors. LEU fuels are viewed as inferior to HEU fuels for military applications due to the need for a larger reactor core to achieve similar performance. In fact, Russia is planning to increase the enrichment of the HEU fuel in its next generation of naval reactors to extend core life and deployment time, to help maintain its sea presence despite downsizing its fleet.

The United States could take four steps to encourage Russia to convert its naval fleet to LEU fuel: 1) Convert the U.S. nuclear naval fleet to LEU fuel; 2) Help convert the Russian icebreaker fleet to LEU; 3) Help commercialize Russia's prototype, LEU-fueled, floating nuclear power plants; 4) Establish a joint research and assistance program to build a new generation of LEU naval reactors for both countries. It is unclear, however, if even these steps would be sufficient, given that Russia perceives no proliferation nor nuclear security risk from using HEU fuel in naval reactors. Russia's confidence stems from its definition of "proliferation resistance" and its significant improvements in nuclear security over the last two decades, funded substantially by the U.S. MPC&A program.

The reported backlog of decommissioned submarines waiting for dismantlement is not an obstacle to the potential conversion to LEU since there are only about 10 to 15 vessels left, with the completion of dismantlement estimated for the year 2015. But Russia does rely on a steady stream of spent naval HEU fuel to help keep the Mayak reprocessing facility operational, directly benefiting some government officials. Any potential conversion of military vessels to LEU fuel could jeopardize operation of the Mayak plant or necessitate costly modifications to handle the new LEU fuel.

Finally, a variant of the prototype RITM-200 reactor being developed by OKBM for the new generation of Russian icebreakers will likely power future Russian nuclear military vessels. RITM-200 will have a cassette-type core with various active zones able to accommodate fuels of varied enrichment. Even if the reactor uses LEU fuel in icebreakers, it might well use HEU fuel in military vessels.

## Conclusions

My findings indicate that phasing out HEU use in Russian nuclear military propulsion would be one of the toughest endeavors in the overall global HEU phase-out. Unlike most worldwide uses of HEU, where justifications for the use of HEU can be assessed, debated, and countered, this is not so in case of Russian naval propulsion. The secrecy surrounding the technical characteristics of military naval reactors blocks information on the fuels currently used, and hinders research and development of operationally acceptable replacement LEU fuels. Unlike research reactors, where an outside country such as the United States can suggest solutions for replacing HEU fuel based on the exact characteristics of the reactors in question, this is not possible for naval military reactors. Therefore, any serious conversion efforts must be explored by Russia itself, although the United States potentially could help by providing funding. However, a strong push for the conversion of Russian military naval reactors is not currently a priority for the United States because of the myriad technical and political obstacles, and other more feasible and obviously beneficial conversion projects.

Still, the United States could take steps to encourage Russia to research and then pursue conversion of its naval fleet to LEU fuel, by pursuing the following policies. The United States should seriously consider converting its own Navy's nuclear fleet to LEU. Before this can happen though, the U.S. Navy must permit outside experts to assess rigorously its claims that conversion to LEU would degrade operational characteristics and increase ship size. If this is indeed so, Congress could provide funding to attempt to overcome these problems.

The United States could also help Russia convert its existing icebreaker fleet to LEU, including jointly developing a new, LEU-fueled propulsion reactor. In addition, the United States could provide funding and expertise to help commercialize Russia's venture into LEU-fueled, barge-mounted power reactors. While these policies could not guarantee conversion of Russian naval reactors, they would at least demonstrate the U.S. commitment to the cause and the feasibility of LEU reactors at sea. Moreover, the United States should reiterate that while the MPC&A program was and is a success, it is not a substitute for further reducing security risks by minimizing the non-weapons uses of HEU.

Russia's government and relevant international organizations could also help promote conversion of naval reactors to LEU fuel. The Russian Ministry of Defense should commission a study focused on present and future military naval requirements and goals, and what operational characteristics their nuclear vessels would need to have, now and in the future. As a part of this study, the Ministry of Defense should seriously consider LEU use in future nuclear-powered military vessels.

The Russian government should consider modernizing and upgrading the Mayak reprocessing facility so that it could reprocess spent LEU fuels from propulsion reactors, as it does with spent HEU fuel, or develop an alternative disposition pathway, so that the back-end of the fuel cycle is not an obstacle to converting naval reactors.

The UN, IAEA, parties to the Treaty on the Non-Proliferation of Nuclear Weapons (NPT), and other actors and organizations should work to close the loophole in the proposed fissile materials production cut-off treaty (FMCT), which allows the production of HEU for use in propulsion reactors. If successful, such action could put normative pressure on Russia to design future naval military reactors to use LEU fuel, even though it has a large stockpile of HEU.

The following are my recommendations for future research. On the technical front, and as secrecy would allow, significantly more research is needed on the potential feasibility of converting to LEU fuel Russia's existing naval reactors:

the VM-4S, VM-4SG, VM-4A, VM-5, KN-3, and OK-650. In addition, more research is required on Russia's prototype RITM-200 reactor for icebreakers, and its potential to use LEU fuel also to power naval reactors, including how to minimize any resulting performance penalties, such as noise. Finally, a detailed examination of present, planned, and potential LEU fuels for naval propulsion would be of great value.

The potential conversion of Russian nuclear military propulsion reactors confronts major technical, economic, and especially political challenges. But, these obstacles are not necessarily insurmountable. The steps and recommendations outlined in this chapter would establish a useful groundwork for future conversion efforts to phase out the worldwide use of HEU fuel in civilian and military naval propulsion.

## Notes

1 The proportion is based on the estimate of global HEU commerce in the first chapter of this book. See also, Ole Reistad and Styrkaar Hustveit, "HEU Fuel Cycle Inventories and Progress on Global Minimization," *The Nonproliferation Review* 15,2 (2008): 265–287.

2 Morten Bremer Maerli, "Components of Naval Nuclear Fuel Transparency," *NATO-EAPC Fellowship Report* (Jan 2002).

3 Ole Reistad, Morten Bremer, and Nils Bohmer, "Russian Nuclear Fuel and Reactors," *The Nonproliferation Review* 12, 1 (2005): 163–197.

4 Reistad and Hustveit, "HEU Fuel Cycle Inventories," 265–287.

5 Nils Bøhmer, Aleksandr Nikitin, Igor Kudrik, Thomas Nilsen, Andrey Zolotkov, and Michael H. McGovern, "The Arctic Nuclear Challenge," *Bellona Report Volume 3* (The Bellona Foundation: 2001).

6 Pavel Podvig, "Consolidation of Fissile Materials in the Russian Nuclear Complex," Proceedings of the 49th Annual Meeting of the Institute of Nuclear Materials Management, 2008.

7 "Reducing and Eliminating Nuclear Weapons: Country Perspectives on the Challenges of Nuclear Disarmament," International Panel on Fissile Materials, 2010.

8 Maerli, "Components of Naval Nuclear Fuel Transparency."

9 Hui Zhang, "Preventing Nuclear Terrorism: Reducing the Danger of Highly Enriched Uranium," Paper presented at the Institute of Nuclear Materials Management 44th Annual Meeting, 2005; "Russia: DOE MPC&A Program," Nuclear Threat Initiative, Inc., http://www.nti.org/db/nisprofs/russia/forasst/doe/mpca.htm (accessed on 10 April 2011).

10 Chunyan Ma and Frank von Hippel, "Ending the Production of Highly Enriched Uranium for Naval Reactors," *The Nonproliferation Review* (Spring 2001): 86–101; "SSBN Borei Class Nuclear Missile Submarine, Russia," http://www.naval-technology.com/projects/borei-class/ (accessed on 27 September 2012). See also, Christine Egnatuk, "Russia: Icebreaker Ships and Floating Reactors," in this volume.

11 N.S. Khlopkin, "Prospects for the Development of Propulsion Nuclear Power Systems," *Atomnaya Energiya* 103, 4 (2007): 211–215.

12 "Universal Nuclear Icebreaker of the New Generation," OKBM Afrikantov, http://www.okbm.nnov.ru/russian/universalicebreaker (accessed on 27 September 2012).

13 P. Podvig, Skype interview by author, Geneva, Switzerland, 10 February 2011.

14 A. Nikitin, Skype interview by author, Oslo, Norway, 17 February 2011. He is the first expert to report that some zones in the cores of Russian naval reactors are enriched above 45%.

15 Podvig, interview.
16 A. Dmitriev, interview by author, Moscow, Russia, 28 February 2011.
17 Nikitin, interview.
18 Ibid.
19 Podvig, interview.
20 Ibid.
21 Nikitin, interview.
22 Dmitriev, interview.
23 V. Kuznetsov, interview by C. Egnatuk, 6 January 2011.
24 Dmitriev, interview.
25 Nikitin, interview.
26 Podvig, interview.
27 Podvig, interview.
28 Kuznetson, interview. The first two Borei class submarines – the Yury Dolgoruky and Alexander Nevsky – were expected to enter service in summer 2012. Gleb Bryanski, "Russia to get stronger nuclear navy, Putin says," *Reuters*, 30 July 2012, www.chicagotribune.com/news/sns-rt-us-russia-putin-navybre86t0zg-20120730,0,4822138.story.
29 Nikitin, interview.
30 See Egnatuk chapter in this volume.
31 Anatoli C. Diakov, Alexander M. Dmitriev, Jungmin Kang, Alexey M. Shuvayev, and Frank N. von Hippel, "Feasibility of Converting Russian Icebreaker Reactors from HEU to LEU Fuels," *Science and Global Security* 14 (2006): 33–48.
32 Dmitriev, interview.

# Part VI

# Ahead of the Curve

# 14 Space Reactors

*R. Blake Messer*

This chapter explores policy solutions to phase out the use of highly enriched uranium (HEU) from use in space nuclear reactors (SNRs). I begin by conducting a literature review to identify the key policy considerations relating to the use of HEU in SNRs. The rest of the chapter examines those considerations.

No SNR projects have been launched since 1988, nor are any being seriously considered for the near future in the United States or elsewhere. Accordingly, global policymakers today have a unique window of opportunity to establish SNR policy to align with the emerging norm of phasing out non-weapons HEU. Policy changes will become much more costly, politically and financially, once governments start seriously considering SNR projects again.

SNRs are not the highest priority for the HEU phase-out mission, but the use of weapons-usable uranium in SNRs is still a legitimate concern for nuclear security. Even if SNRs can be relied on to safely launch, operate, and retire in outer space, the substantive policy issue is that HEU phase-out must be nondiscriminatory in order to be successful. States without nuclear weapons will have stronger diplomatic grounds to resist global nonproliferation norms if the states with such weapons fail to reinforce these norms concretely and by example. Seemingly benign exceptions could compromise the global phase-out of HEU. In short, security policies that are not adopted domestically will be difficult to push internationally.

There are also domestic political barriers to eliminating HEU from use in space nuclear reactors. SNRs present the most sophisticated technical challenge to a global HEU phase-out because HEU's high energy density and low radioactivity make it a uniquely attractive fuel for space power and propulsion. Further, the tenuous nature of governmental commitments to space exploration and research makes scientists resistant to policy mandates that increase the funding or time required to bring projects to completion. Reevaluating the use of HEU could thus make a project vulnerable to discretionary budget cuts. HEU also has tradition on its side, because all SNRs ever launched – 35 by the Soviet Union and one by the United States – have used weapons-grade HEU fuel of 90 percent enrichment or higher.[1]

Policy solutions that phase out HEU from use in SNRs could be expensive and therefore provoke resistance from government researchers and scientists.

This is mainly because LEU fuel typically requires more mass to produce the same energy as HEU fuel. Conversion from HEU to LEU fuel could thus increase costs for space reactor missions significantly more than for other applications of nuclear reactors because of the premium on minimizing weight in space missions.

Despite these technical and political challenges, pursuing conversion via policy incentives is not impossible and would not necessarily require any sacrifice of goals by the United States or any foreign government intent on launching space reactors. My research found that no serious effort has been made to assess the costs of LEU conversion for SNRs. Accordingly, no analysis currently shows that adopting LEU would be prohibitively expensive for SNR projects. I make several recommendations at the end of this chapter that could facilitate an international shift toward LEU fuel in SNRs.

My research also investigated a UN resolution that appears to ban LEU fuel in SNRs. I found that the United States does not, in fact, consider the use of LEU in SNRs to violate the UN resolution. Officials of the U.S. Department of Energy (DOE) Office of Nuclear Energy (NE) involved in the resolution stated that it was drafted primarily to prevent countries from using plutonium in SNRs and was not intended to preclude the use of LEU or other potentially safe fuels that at the time had yet to be envisioned for SNR applications. I also found that no space reactor launches are currently being seriously considered in the United States, Europe, or Japan, nor is there evidence that Russia or China have any serious timetables for an SNR launch.

The next section of this chapter documents the body of known policy literature on the use of HEU in SNRs, including historical uses of HEU in SNRs by the United States and Soviet Union, domestic and international regulatory policies for SNRs, and current developments in SNRs. The third section discusses methodologies used in this research, primarily interviews with experts and officials about SNRs and the policies that govern them, combined with secondary research. The fourth section contains a comprehensive discussion of the findings of my research. Finally, the last section summarizes key findings and ends with conclusions and policy recommendations.

## Literature Review

Historically, using HEU fuel to power SNR systems has been standard practice. The United States and Soviet Union used weapons-grade HEU for each of the 36 SNRs launched to date.[2] Traditionally, SNRs have also been known as "nuclear electric propulsion" (NEP) systems, which convert energy from a nuclear reactor into electricity that is then used to ionize a gas for propulsion. They are distinct from "nuclear thermal propulsion" (NTP) systems, which use the decay of nuclear materials to heat and eject a propellant directly, a method that is favored where spacecraft need to move faster. The divide between these two terms is no longer rigid as the technologies are now somewhat integrated.[3]

### *U.S. Space Reactors*

In April 1965, the United States launched into orbit the world's first SNR, the SNAP-10A. The reactor remained operational for 43 days before a component failure on the satellite, unrelated to the reactor system, forced it to shut down. The satellite was subsequently retired by being boosted into high orbit where it remains today.[4] The United States has not since launched another SNR.

The U.S. National Aeronautics and Space Administration (NASA) had earlier begun work in 1958 (the year it was created) on Project ORION, which notably proposed to use a series of nuclear explosions to propel a spacecraft into orbit. But ORION was cancelled in 1963, when the U.S. signed on to the Partial Test Ban Treaty, which prohibited such a project. From 1961 to 1973, the Nuclear Energy for Rocket Vehicle Application (NERVA) program developed and tested rockets envisioned for interplanetary travel that utilized nuclear reactors as the main heat source. As part of the NERVA program, 21 reactor tests were run at a Nevada test site, but no such reactor was launched into space and the program was discontinued due to budget constraints.[5] Project Prometheus began in 2003 as an NEP reactor development effort between the Department of Defense (DOD) and other U.S. government agencies, with the aim of shortening flight times to Mars or the outer planets.[6] But Prometheus was cancelled in 2005, also due to budget constraints.[7]

For all proposed launches of nuclear-powered systems, the United States uses a safety review procedure known as the Nuclear Launch Safety Approval (NLSA). The NLSA requires participation from four agencies: DOE, which governs the development and use of nuclear technologies and submits safety analysis reports for specific mission designs; DOD, which controls launch facilities; NASA, which plans and designs space missions; and the White House Office of Science and Technology Policy (OSTP), which may authorize a launch or pass it to the president for final approval. Final safety analyses are conducted by the Interagency Nuclear Safety Review Panel before a launch authorization request is sent to OSTP.[8]

### *Soviet Space Reactors*

The Soviet Union launched into space 35 known HEU-fueled SNRs between 1969 and 1988 (see Table 14.1). All but two of these were used in Radar Ocean Reconnaissance Satellites (RORSATs), which were built to track U.S. naval positions.[9] The remaining two were TOPAZ-1 reactors – the last model launched by the Soviet Union – to test the reactors, which were fueled with 96%-enriched uranium dioxide fuel pellets, and their electric engines.[10]

### *International SNR fuel policy*

In 1992, the UN General Assembly passed the "Principles Relevant to the Use of Nuclear Power Sources in Outer Space," a resolution that makes stipulations for "nuclear power sources in outer space devoted to the generation of electric power

*Table 14.1* Soviet Union's Space Nuclear Reactor Launches

| *Date* | *Spacecraft* | *Lifetime* |
|---|---|---|
| 1967-Dec-27 | Cosmos 198 | 1 day |
| 1968-Mar-22 | Cosmos 209 | 1 day |
| 1970-Oct-3 | Cosmos 367 | 1 day |
| 1971-Apr-1 | Cosmos 402 | 1 day |
| 1971-Dec-25 | Cosmos 469 | 9 days |
| 1972-Aug-21 | Cosmos 516 | 32 days |
| 1973-Dec-27 | Cosmos 626 | 45 days |
| 1974-May-15 | Cosmos 651 | 71 days |
| 1974-May-17 | Cosmos 654 | 74 days |
| 1975-Apr-2 | Cosmos 723 | 43 days |
| 1975-Apr-7 | Cosmos 724 | 65 days |
| 1975-Dec-12 | Cosmos 785 | 1 day |
| 1976-Oct-17 | Cosmos 860 | 24 days |
| 1976-Oct-21 | Cosmos 861 | 60 days |
| 1977-Sep-16 | Cosmos 952 | 21 days |
| 1977-Sep-18 | Cosmos 954 | 43 days (reentered) |
| 1980-Apr-29 | Cosmos 1176 | 134 days |
| 1981-Mar-5 | Cosmos 1249 | 105 days |
| 1981-Apr-21 | Cosmos 1266 | 8 days |
| 1981-Aug-24 | Cosmos 1299 | 12 days |
| 1982-May-14 | Cosmos 1365 | 135 days |
| 1982-Jun-1 | Cosmos 1372 | 70 days |
| 1982-Aug-30 | Cosmos 1402 | 120 days (reentered) |
| 1982-Oct-2 | Cosmos 1412 | 39 days |
| 1984-Jun-29 | Cosmos 1579 | 90 days |
| 1984-Oct-31 | Cosmos 1607 | 93 days |
| 1985-Aug-1 | Cosmos 1670 | 83 days |
| 1985-Aug-23 | Cosmos 1677 | 60 days |
| 1986-Mar-21 | Cosmos 1736 | 92 days |
| 1986-Aug-20 | Cosmos 1771 | 56 days |
| 1987-Feb-1 | Cosmos 1818 | ±6 months |
| 1987-Jun-18 | Cosmos 1860 | 40 days |
| 1987-Jul-10 | Cosmos 1867 | ±1 year |
| 1987-Dec-12 | Cosmos 1900 | 124 days |
| 1988-Mar-14 | Cosmos 1932 | 66 days |

Source: Steven Aftergood, "Background on Space Nuclear Power," *Science & Global Security* 1 (1998): 97.

on board space objects for non-propulsive purposes," which excludes most space nuclear reactor missions envisioned for civilian applications.[11] It stipulates that "[space] nuclear reactors shall use only highly enriched uranium 235 as fuel."[12] Though nonbinding, this resolution seems in practice to be taken seriously by spacecraft engineers considering alternative fuels.[13]

The United States signed the Principles document, but submitted a letter of reservation exempting its space programs from the requirements pertaining to reactor fuel, stating:

On some points ... it remains our view that the principles related to safe use of nuclear power sources in outer space do not yet contain the clarity and technical validity appropriate to guide safe use of nuclear power sources in outer space. The United States has an approach on these points which it considers to be technically clearer and more valid and has a history of demonstrated safe and successful application of nuclear power sources. We will continue to apply that approach.[14]

### *Recent Developments*

Nuclear reactors are now envisioned as a future power source on the surface of other planets, moons, or asteroids.[15] Former U.S. President George W. Bush's "Vision for Space Exploration" aimed to conduct lunar surface missions in the decade starting in 2020, and a Mars surface mission in the following decade. NASA's website claims that "the most mass-efficient means of providing high power for surface missions is through the use of nuclear fission systems."[16] In 2005, DOE signaled that it would allocate 20 metric tons of weapons-grade HEU for space reactors and research reactors.[17] In Russia, in 2007, then-President Dmitry Medvedev made public statements about his desire to "find room in the budget" for Russia to use a 1-megawatt space nuclear reactor to send a human to Mars, but no such project seems currently to be under construction or immediate consideration.[18]

There has been only one serious academic exploration of using LEU to fuel a space reactor. A 2009 paper by Nagata and co-authors, at Japan's Kyushu University, outlines an LEU-fueled SNR design for an interstellar explorer.[19] The reactor discussed in their paper would provide 100 $kW_t$ for 7 years and weigh less than 500kg.[20]

## Methodology

My primary sources comprised mainly interviews with experts, researchers, and officials from DOE and the European Science Foundation (ESF). These interviews were conducted either in person, via email, or by phone. In-person interviews were also conducted at the conference on Nuclear and Emerging Technologies for Space, held in Albuquerque, New Mexico, in February 2011.

My first goal was to document any existing analyses of the penalties of using LEU rather than HEU fuel in space reactors. I discussed this topic with relevant experts and sought documents that assessed whatever sacrifices in weight, range, and/or cost could be associated with conversion of SNRs to LEU. I also communicated via email with Dr. Frank von Hippel, who was Assistant Director for National Security at OSTP when it requested that DOE assess the weight penalties for LEU conversion.

Second, I investigated what motivated Nagata, et al., to formulate an SNR design based on LEU fuel, as well as the prospects for their design to advance toward implementation. I communicated via email with Hideki Nakashima, a co-author of that study. Third, I examined whether DOE has specific plans for the 20 metric tons of HEU that it withdrew from U.S. weapon stocks for space reactors and research reactors. I investigated this question by communicating via email and phone with DOE officials. Fourth, I explored the origins and motivations of the UN resolution that prohibits using fuel other than HEU for space reactors to power electrical systems. I investigated this question by interviewing DOE officials who were involved in the UN deliberations. Fifth, I investigated SNR projects in other countries, seeking out experts and government officials from the EU, Japan, Russia, and China.

The following experts and government officials spoke with me about these research questions:

- Alice Caponiti – Program Director for Infrastructure Capabilities, Office of Space and Defense Power Systems, Office of Nuclear Energy, DOE.
- Dr. Frank von Hippel – Co-Director of the Program on Science and Global Security, Princeton University, and former Assistant Director for National Security in the White House Office of Science and Technology Policy.
- Hideki Nakashima – Professor of Energy System Science in the School of Engineering Sciences, Kyushu University, Japan.
- Dr. Jean-Claude Worms – Head of the Space Science Unit, European Science Foundation.
- Robert Lange – Deputy Assistant Secretary for Business and Technical Support, Office of Nuclear Energy, DOE.
- Ryan Bechtel – Space Reactor Technology Program Manager, Office of Space Defense Power Systems, DOE.
- Dr. Ralph McNutt – Johns Hopkins University Applied Physics Laboratory.
- Wade Carroll – Deputy Director of the Office of Space and Defense Power Systems, DOE.

My research faced some constraints. I had difficulty obtaining responses from government officials in Russia and so was unable to confirm whether that country is seriously considering an SNR launch. In addition, U.S. law prohibited me from contacting the appropriate officials in China, because the Chinese Academy of Space Technology (CAST) has been subject to U.S. sanctions since 1993 as a subsidiary of the China Aerospace Science and Technology Corporation (CASC), accused of missile-related sales to Pakistan.[21]

## Findings

### *Penalty of Using LEU in SNRs is Unknown*

In 1994, a DOE study assessed the weight penalty for LEU conversion. It found that converting the 40 $kW_e$ S-Prime reactor from HEU to LEU fuel would increase its weight from 2.7 tonnes to 4.2 tonnes, a 56 percent increase. But Dr. Frank von Hippel, who requested the study when he was Assistant Director for National

Security at OSTP, convincingly argues that the study's estimates exaggerate the penalty. In 2007, he gave a presentation to the American Nuclear Society's Space Nuclear Conference, arguing that the DOE study failed to optimize the LEU design for the reactor or spacecraft.[22] Dr. von Hippel notes that the DOE study kept constant the reactor core's height, while increasing the diameter of the cross-section by 75 percent to account for the higher volume of LEU. This biased upward the weight estimates because an optimized LEU core would have both a larger height and diameter compared to an HEU core, which would reduce the weight compared to the design assumed by DOE. Dr. von Hippel says that even DOE's estimate of a 56 percent weight increase is "not that extraordinary," although he concedes it is a "weight gain that spacecraft designers would prefer to use for something else."[23]

DOE's study is also somewhat outdated and incomplete. For example, current and prospective SNR designs may differ considerably from those envisioned in 1994. In addition, the United States has never assessed rigorously the financial penalty of using LEU rather than HEU to fuel an SNR. The Japanese study by Nagata et al. may be the best indicator of the cost of a realistic LEU-fueled SNR system, but more research would be needed to vet that design for feasibility. Those authors estimated that their design would weigh less than 500kg while generating 100kW$_t$, but they do not compare the weight to that of an HEU design with equivalent power output. Further study is therefore required to accurately assess the cost of converting SNRs to LEU.

### *No Launches Currently Planned*

I investigated whether the United States, EU, Japan, Russia, or China are actively considering SNRs. Officials at DOE told me that no U.S. projects are under serious consideration in the near term. But they did report that research is being conducted on a 40 kW$_e$ SNR system as a potential fission power source on the surface of the moon, Mars, or an asteroid. Robert Lange, DOE's Deputy Assistant Secretary for Business and Technical Support, characterized it as a "low-level development effort [that is] not yet funded for construction." Electrical tests of the system will take place in 2012 at the Glenn Research Center, in Ohio, but officials stated that there are no identified missions for the reactor at this time and that they had no timeline for when a possible launch might occur. In November 2012, Los Alamos National Laboratory announced that it had conducted the first demonstration of a space nuclear reactor system in the United States since 1965 – likely the same project identified above – producing 24 watts of electricity at the Nevada National Security Site's Device Assembly Facility near Las Vegas.[24]

Professor Hideki Nakashima, one of the Kyushu University researchers who published a design for an LEU-fueled SNR system, told me that "unfortunately, space nuclear reactor [sic] is not seriously considered by the Japanese government." As for the EU and the European Space Agency (ESA), Dr. Jean-Claude Worms, head of the Space Sciences Unit at the European Science Foundation, reported the following in an interview:

> Unfortunately there is currently no serious planning at EU level concerning [space] nuclear reactors. The topic has been debated for years in various fora, technical and political, but its sensitivity in some EU or ESA member states restrained the political discussion at national level (France, UK, Italy) or within ESA and the science/technology community for debating technical options ... The current (7th) Framework Program of the European Commission does not address space nuclear propulsion, and it is probable that the next Framework Program (2014–2020) will not address it either. The only elements that might be included there are radio-isotopic devices for spacecraft energy and heating, under the heading "critical technologies for European non-dependence." Even this remains to be seen, although I have recently noticed an increased interest in Europe for RTGs, RHUs, etc.[25]

Based on these interviews, it appears that both the United States and EU have researched and discussed launching SNRs in the future, but neither currently has a timetable for a launch. Nor is Japan currently considering SNR projects in the near term. Regarding DOE's plans for the 20 metric tons of HEU reserved for space reactors and research reactors, Ryan Bechtel in the Office of Space and Defense stated only that "the DOE has reserved a sufficient amount of nuclear fuel to meet NASA's space reactor mission planning set needs through 2040." Another DOE official who requested anonymity stated that the specific breakdown of the 20MT is classified.

### *UN Principles do Not Prohibit LEU in SNRs*

My research revealed that the United States does not consider the use of LEU in SNRs to violate the UN resolution on the "Principles on the Use of Nuclear Power Sources in Outer Space." Wade Carroll, Deputy Director of the Office of Space and Defense Power Systems at the U.S. Department of Energy, and Robert Lange, Deputy Assistant Secretary for Business and Technical Support of the Office of Nuclear Energy at the U.S. Department of Energy, stated that the UN's requirement for HEU fuel was drafted primarily to prevent the use of plutonium in SNRs on environment grounds. They do not believe it was intended to preclude the use of LEU or other potential fuels that in 1992 had not yet been envisioned for SNR applications. They also confirmed that the UN resolution would likely not preclude the United States from using LEU fuel in an SNR. But officials in DOE's Office of Nuclear Energy also stated that at the moment they would not necessarily support amending the UN resolution to explicitly allow for LEU fuel because they "currently have no projects under development that would require that change."

### *U.S. DOE Has Not Seriously Considered LEU in SNRs*

Ryan Bechtel, the program manager for Space Reactor Technology at DOE's Office of Space and Defense, reported that DOE has not yet seriously considered LEU-fueled SNRs:

> Only HEU space reactor systems have been considered because of total system mass considerations. Mass is a limiting factor on all space missions. Though it may be less of an issue for surface power systems, it still has a major impact on launch costs. HEU systems (fuel and the corresponding shielding, reflectors, heat transport and energy conversion systems, etc.) are inherently lower mass and volume than LEU systems. If there is any single case for which a higher mass, LEU system may be deemed acceptable from a mass penalty perspective, then it would be a separate development effort than what would be needed for all other applications. Such a separate development and implementation effort would be a costly approach.[26]

Though DOE has not considered LEU systems in past or current development efforts, it is contemplating heavier launch vehicle systems for the future that could more easily accommodate an LEU-fueled SNR. One DOE official who requested anonymity noted that in light of the substantial development costs for a heavy-lift system, "the cost to develop an LEU" reactor system would be "a small difference relative to the costs of developing a new launch vehicle."

### *Challenges of Phasing out HEU in SNRs*

SNRs present the least urgent but most complicated technical challenge to a global HEU phase-out, because HEU's high energy density and low radioactivity make it a uniquely attractive fuel for space power and propulsion. In addition, the political challenges of converting SNRs from HEU to LEU are almost as daunting as the technical ones. At the 2011 conference on Nuclear and Emerging Technologies for Space, a DOE contractor told me that "the best way to kill space nuclear research would be to mandate LEU for space nuclear reactors." His assessment is understandable. The tenuous nature of Congressional commitments to space exploration and research makes U.S. government scientists resistant to policy mandates that would increase the funding or time required to bring projects to completion. Both factors tend to flag government research for discretionary spending cuts in the Congressional budget process. Scientists in other countries face similar budgetary pressures. Meanwhile, policymakers lack sufficient information about the penalties of converting SNRs to LEU fuel to formulate a policy. To develop that knowledge base in the United States, Congress or the OSTP probably would need to mandate that DOE conduct a realistic assessment, given that the bureaucracy appears averse to initiating such a study on its own initiative.

### *Summary of Findings*

My research found that no government has made a serious effort to assess the penalties of using LEU fuel in SNRs. As a result, no analysis currently shows that converting to LEU would be prohibitively expensive for all SNR projects. My research also revealed that the United States does not consider the use of LEU in

SNRs to violate the 1992 UN resolution that ostensibly requires HEU fuel for space nuclear power, because the resolution's actual intent was to ban plutonium fuel on environmental grounds. I also found that no SNR projects are currently being considered on a concrete timetable by the United States, EU, or Japan, nor is there evidence of such plans in Russia or China.

## Conclusions and Policy Recommendations

SNRs are not the most proliferation-prone use of HEU, nor the most cost-effective target for conversion to LEU. Nevertheless, several small measures could pave the road for the potential phase-out of HEU in future SNRs. Specifically, the U.S. government and the UN's Office of Outer Space Affairs could take the following steps.

### *Recommendations for the U.S. government*

The White House's OSTP should request that DOE work with the RERTR program to assess rigorously the penalties of SNRs that are optimized for LEU fuel, compared to those optimized for HEU fuel. In the meantime, the RERTR program should communicate proactively with DOE's Office of Nuclear Energy to ensure that high-density LEU fuel is taken into account in any conversion cost estimates. To support the emerging international norm of avoiding HEU in non-weapons applications of nuclear power, Congress should appropriate funds to the DOE for development of LEU-fueled SNR designs. If Congress ultimately deems the nonproliferation objective sufficiently worthy, it also should provide funding to offset the cost increases from using LEU fuel in future SNRs.

### *Recommendations for the United Nations*

I recommend that the United Nations revise the 1992 UN Principles on the Use of Nuclear Power in Outer Space to explicitly allow for LEU fuel in SNRs. Although no individual country has an urgent incentive to amend the UN resolution, it is important to remove any impression that LEU fuel is banned in space reactors. The current language sends a signal that is inconsistent with international nonproliferation norms.

## Notes

1 Frank von Hippel, "A Comprehensive Approach to Elimination of Highly-Enriched-Uranium From All Nuclear-Reactor Fuel Cycles," *Science & Global Security* 12, 3 (2004): 137.
2 von Hippel, "A Comprehensive Approach to HEU Elimination," 137.
3 Nathan Voegeli, "Space Nuclear Reactors: History and Emerging Policy Issues," *The Nonproliferation Review* 14, 1 (March 2007): 163–175.
4 William C. Potter, "Nuclear Terrorism and the Global Politics of Civilian HEU Elimination," *The Nonproliferation Review* 15, 2 (2008): 135–158.

5 Voegeli, "Space Nuclear Reactors: History and Emerging Policy Issues," 163–175.
6 William C. Potter, "Nuclear Terrorism and the Global Politics of Civilian HEU Elimination," 135–158. von Hippel. "A Comprehensive Approach to HEU Elimination," 137.
7 "Nuclear Reactors for Space," http://www.world-nuclear.org/info/inf82.html (accessed on 27 September 2012). Randall Taylor, "Prometheus Project Final Report," NASA, http://hdl.handle.net/2014/38185 (accessed on 13 November 2012).
8 Voegeli, "Space Nuclear Reactors: History and Emerging Policy Issues," 163–175. Steven A. Mirmina and David J. Den Herder, "Nuclear Power Sources and Future Space Exploration," *Chicago Journal of International Law* 6 (2006): 149.
9 Voegeli, "Space Nuclear Reactors: History and Emerging Policy Issues," 163–175.
10 Voegeli, "Space Nuclear Reactors: History and Emerging Policy Issues," 163–175.
11 Voegeli, "Space Nuclear Reactors: History and Emerging Policy Issues," 163–175.
12 United Nations General Assembly, "Principles Relevant to the Use of Nuclear Power Sources in Outer Space," A/RES/47/68, 14 December 1992, http://www.un.org/documents/ga/res/47/a47r068.htm (accessed on 27 September 2012).
13 Ralph L. McNutt, "Phase II Final Report: A Realistic Interstellar Explorer," NASA Institute for Advanced Concepts, 14 October 2003, http://www.niac.usra.edu/files/studies/final_report/393McNutt.pdf (accessed on 27 September 2012). Also see Mirmina and den Herder, "Nuclear Power Sources and Future Space Exploration," 149.
14 Kenneth Hodgkins, Statement to the United Nations General Assembly, 28 October 1992, http://www.state.gov/documents/organization/65951.pdf (accessed on 27 September 2012).
15 NASA, "Fission Surface Power," http://www.grc.nasa.gov/WWW/TECB/fsp.htm (accessed on 27 September 2012).
16 NASA, "Fission Surface Power."
17 Alexander Glaser and Zia Mian, "Fissile Material Stockpiles and Production," *Science & Global Security* 16, 3 (2008): 59. Potter, "Nuclear Terrorism and the Global Politics of Civilian HEU Elimination," 148, refers to this as entirely for space reactors, apparently incorrectly.
18 "Russia Proposes Nuclear Spaceship – CBS News," http://www.cbsnews.com/stories/2009/10/29/tech/main5449049.shtml (accessed on 27 September 2012).
19 Hidetaka Nagata, Takaaki Nishiyama, and Hideki Nakashima. "Proposal of Space Reactor for Nuclear Electric Propulsion System," *Transactions of Space Technology Japan* 7 (2009).
20 Nagata, et al., "Proposal of Space Reactor for Nuclear Electric Propulsion System."
21 Nuclear Threat Initiative, "China Academy of Space Technology (CAST)," http://www.nti.org/facilities/56/ (accessed on 27 September 2012).
22 Frank von Hippel, "Two Policy Issues: 1) Safety of RTG Earth Flybys 2) HEU vs. LEU space-reactor fuel," presented at American Nuclear Society Space Nuclear Conference, Boston, MA, 27 June 2007.
23 von Hippel, "Two Policy Issues."
24 Los Alamos National Laboratory, "Joint DOE and NASA team demonstrates simple, robust fission reactor prototype," 26 November 2012, http://www.lanl.gov/newsroom/news-releases/2012/November/11.26-space-travel.php.
25 See, for example, Major S. Chahal, "European Space Nuclear Power Programme: UK Activities," UK Space Agency, 9 February 2012, http://www.oosa.unvienna.org/pdf/pres/stsc2012/tech-18E.pdf (accessed on 13 November 2012).
26 Interview with author, 21 January 2011.

# Part VII

# Conclusion

# 15 Achieving a Global HEU Phase-Out

*Alan J. Kuperman*

It will be no easy task to eliminate worldwide non-weapons use of highly enriched uranium. The opponents are powerful and predictable, judging from their past resistance to even limited attempts to reduce HEU commerce. Many scientists and engineers will favor continued HEU use because it is efficient and familiar – and because their main concerns are cost and performance, rather than global security. They initially resisted conversion of low-power research reactors to LEU fuel, then conversion of high-performance reactors, and so will probably also oppose any further expansion in scope of an HEU phase-out. Many in the private sector will favor HEU use wherever it may provide efficiencies that offer competitive advantage. The world's leading medical-isotope producers for years resisted pleas to phase out HEU from their manufacturing processes. So, it can be anticipated that firms in other sectors will likewise demand HEU for future commercial ventures – whether space reactors, nuclear-propelled ships, or other applications. The militaries of the United States and Russia will insist that continued use of HEU fuel for naval propulsion is essential to national security, because any reduction in performance would leave them vulnerable to potential adversaries. As a result, other existing and emerging naval nuclear powers will demand the same right – on similar grounds or simply because they oppose the double-standard. Finally, some countries will insist on using HEU for non-weapons purposes because it provides the ideal cover for a weapons program, or at least maintaining the option. Iran notably declared in June 2012 that it had commenced a program to produce nuclear-propelled submarines.[1]

Fortunately, the case studies in this book highlight lessons for overcoming such powerful resistance. First, it is important for at least one country to take the lead by setting an example, and the United States may need to continue playing that role. The U.S. government in 1978 created the international program to convert reactors, in 1986 ordered the conversion to LEU fuel of the country's licensed research reactors, and more recently committed to convert its remaining high-performance reactors. Such leadership has fostered international cooperation in many ways. It provided the collective good of high-density LEU fuel, which other countries might not have developed on their own but could utilize. It also demonstrated that conversion to LEU fuel was technically feasible at little to no cost in reactor performance. In addition, it showed that the United States was practicing what it preached and was willing to pay the small economic penalty of conversion in order to reduce global security risks. Russian nuclear officials cited in this book say that their decision to explore domestic conversion was driven heavily by the precedent of the United States committing to convert its own high-performance reactors.

Second, it is vital to sustain and expand normative support for HEU minimization. Admittedly, norms are often a weak reed in international politics, but there is increasing evidence for the effectiveness of this one. For example, this volume reports that Russia decided to use LEU fuel in its prototype floating nuclear power plant, and its next generation of nuclear-propelled icebreaker ships, because it envisioned marketing those items abroad. Russian officials say they feared that such exports would be difficult with HEU fuel, even though formally the norm applies only to research reactors and medical-isotope production. This Russian action suggests that expanding the scope of the anti-HEU norm may be easier than some suspect, especially if the norm's traditional scope continues to be observed and enforced.

A third lesson is that minimizing future HEU use will be easier if countries do not start with HEU in the first place, owing to path dependency. For example, China's nuclear-propelled submarines today use LEU fuel – despite the country now having the funds, industrial capacity, and access to raw materials to produce sufficient HEU – probably because its program started with LEU fuel and now is habituated to it. Conversely, Germany's FRM-II reactor continues to use weapons-grade HEU fuel, despite a German government commitment to reduce the fuel's enrichment, because the facility's original design made it technically difficult or expensive to convert. One clear implication is that as additional states pursue nuclear navies, it is important that they start with LEU rather than HEU fuel.

The case studies also illustrate tools that can encourage states to minimize HEU use. Most obvious is the leverage stemming from supply and consumption. Argentina, Belgium, and the Netherlands all agreed to convert their medical-isotope production processes to LEU targets because the United States declared that it would no longer supply HEU for this purpose. No such leverage exists towards Russia, which has its own HEU, so the United States instead plans to utilize the leverage arising from consumption, by means of "preferential procurement." Washington has declared that it will pay higher reimbursement for medical isotopes produced without HEU, aiming either to persuade Russia to convert to LEU targets, or otherwise to reduce Russia's ability to sell its Mo-99 in the country that imports half of the world's production. Either outcome would reduce Russia's use of HEU and thereby reinforce the global norm.

Positive incentives also can promote HEU minimization. For example, South Africa's medical isotope producer traditionally had used HEU targets because the country had a supply of that material and inadequate funding to convert to LEU targets. But soon after the United States provided a $25 million grant, South Africa started producing partially with LEU targets, and now plans to convert its remaining production when foreign markets license the LEU-produced isotope. In Russia, unfortunately, financial incentives perversely encourage HEU use. If a

Russian facility converts from HEU to LEU fuel, worker salaries are cut. Merely contemplating conversion makes a Russian facility more vulnerable to closure, so operators and workers understandably resist. It is essential to get the incentives right by ending practices that penalize workers for minimizing HEU. If Russia cannot afford the extra expense, other states should help.

Because countries typically fear any competitive disadvantage, HEU minimization can also be facilitated by multilateral agreements that assure a level playing field. For example, four states joined in the 2012 quadripartite agreement to eliminate HEU in the production of Mo-99 by 2015. The Netherlands and Belgium, home to two of the world's three largest producers of the medical isotope, had each previously feared suffering a comparative loss if it converted while the other did not. By eliminating the threat of free-riding, the multilateral agreement enabled collective action, just as theory would predict.[2] A similar approach may be necessary to phase out HEU for naval propulsion, given that countries fear a security disadvantage if they convert to LEU fuel while others do not. Russia and the United States, the two largest users of HEU for naval propulsion, ideally would lead such an agreement. But if a bilateral accord cannot be reached, the United States – in light of its substantial military superiority – could boldly declare unilaterally that it intended to convert its nuclear navy to LEU fuel as soon as possible, and then challenge other countries to join in. That is similar to what the United States did for medical-isotope production, by first announcing that it would eschew the use of HEU to produce Mo-99 and then helping forge the quadripartite agreement.

This book's case studies also offer practical insight for implementing a global HEU phase-out. First, it turns out that technical innovation often follows, rather than precedes, a decision to convert. This has implications for today's claims by the United States and Russia that converting their nuclear navies to LEU fuel would sacrifice performance. To the contrary, after France decided to convert its naval propulsion because it did not want to restart HEU production, its designers managed to innovate in ways that actually improved performance while reducing enrichment. Second, scientific collaboration may achieve what diplomacy cannot. For example, Russia's initial progress toward conversion came from interactions with U.S. national laboratories – so-called "lab-to-lab" cooperation – more than a decade before diplomats eventually signed a bilateral agreement to conduct conversion feasibility studies. Similarly, trans-Atlantic scientific collaboration is facilitating development of higher-density UMo LEU fuels, which is a prerequisite for conversion of high-performance reactors.

Admittedly, past experience suggests that the road to a global HEU phase-out will be long and winding. Whether for technical or political reasons, previous efforts to reduce HEU use have required decades in each sector where they have been pursued. For example, the RERTR program was established in 1978 to convert research reactors, but not until 45 years later are the last high-performance reactors projected to be converted. Similarly, after the conversion program was expanded to medical isotope production in the mid-1980s, it will have been 30 years before the major companies are expected to convert in 2015. This illustrates two lessons. First, it is essential to start exploring conversion, or other means of phasing out HEU use, in every sector as soon as possible. For example, even though no space-reactor missions currently are on the drawing board, now is the moment to initiate conversion feasibility studies, so that an informed decision about fuel enrichment can be made when the time inevitably comes for the next mission. Similarly, while it may be too late to convert the next-generation of U.S. and Russian nuclear submarines to LEU fuel because their design stages are already too advanced, it is not too early to start research and development of an LEU-fueled reactor for the follow-on generation of nuclear-propelled vessels, to be deployed a few decades later.

The final lesson is about persistence. Opponents of HEU minimization have tried many times to derail the effort. In the 1980s and early 1990s, for example, they succeeded at eliminating funding for development of higher-density LEU fuels, needed to convert high-performance reactors, and of LEU targets for medical-isotope production. But nuclear-security advocates, inside and outside of government, persisted and in the mid-1990s revived both programs, which are now bearing fruit two decades later. In the same vein, during the first decade of the millennium, Russian opponents blocked conversion of their country's research reactors. But persistent efforts by the U.S. government, Russian supporters, and non-governmental organizations have helped overcome that opposition.

Similar persistence will be needed to broaden the scope of current HEU minimization efforts into a global HEU phase-out. The good news is that the past successes documented in this book suggest the objective is achievable. If sufficient political will can be generated to embrace the principle of eliminating non-weapons use of HEU, then the technical and economic challenges likely can be overcome. Admittedly, there will be times when technical or political progress will stall. At those dark moments, advocates of an HEU phase-out will need to renew their energies by recalling their original motivation. Simply put, so long as HEU commerce persists for non-weapons purposes, the danger of nuclear terrorism will be much greater than it needs to be. That is one risk to humanity that can, and should, be phased out.

## Notes

1 Greg Thielmann and Wyatt Hoffman, "Submarine Nuclear Reactors: A Worsening Proliferation Challenge," Arms Control Association, 26 July 2012.

2 Mancur Olson, *The Logic of Collective Action: Public Goods and the Theory of Groups* (Cambridge, Mass.: Harvard University Press, 1971).

# Index

References in **bold** indicate tables and in *italics* indicate figures.

Printed in the USA/Agawam, MA
November 18, 2013

582097.204